VISIT

This

PLANET

The Andromeda spiral galaxy, Merlin's home—a gravitationally bound collection of more than 300 billion stars located 2.2 million light-years from the Milky Way.

Also by Neil de Grasse Tyson

UNIVERSE DOWN TO EARTH

MERLIN'S TOUR OF THE UNIVERSE

Neil de Grasse Tyson

Illustrations by Stephen J. Tyson

MAIN
STREET
BOOKS

DOUBLEDAY

New York

London

Toronto

Sydney

Auckland

JUST
VISITING
This
PLANET

Merlin Answers More Questions
About Everything Under the Sun,
Moon, and Stars

A MAIN STREET BOOK
PUBLISHED BY DOUBLEDAY

a division of Bantam Doubleday Dell Publishing Group, Inc.
1540 Broadway, New York, New York 10036

DOUBLEDAY, MAIN STREET BOOKS, and the portrayal of a building with a tree
are trademarks of Doubleday, a division of
Bantam Doubleday Dell Publishing Group, Inc.

BOOK DESIGN BY CAROL MALCOLM RUSSO/SIGNET M DESIGN, INC.

Library of Congress Cataloging-in-Publication Data
Tyson, Neil de Grasse.
Just visiting this planet / Neil de Grasse Tyson. — 1st ed.
p.   cm.
Includes index.
1. Astronomy—Miscellanea.   2. Astrophysics—Miscellanea.
I. Title.   QB52.T94   1998
520—dc21   97-47122
CIP

ISBN 0-385-48837-8
Text copyright © 1998 by Neil de Grasse Tyson
Illustrations copyright © 1998 by Stephen J. Tyson

*To all those
who have always looked up,
and to all those
who do not yet know
why they should*

*A*CKNOWLEDGMENTS

I began writing the Merlin column for *Star Date* magazine (then known as the *McDonald Observatory News*) in 1983, and have written it ever since. While the character development of Merlin and the full content of both books derived from the column are original to me, the Q&A column itself was created several years before by David Slavsky, then a graduate student at the University of Texas at Austin and now dean of students at Loyola University in Chicago. Between Slavsky and me, there were short terms of authorship by Jeff Brown and Suzanne Hawley.

During the first few years that I wrote the column, Damond Benningfield, then editor of *Star Date* (and now its executive editor), offered several stylistic insights that have remained with the column. One of them is the fact that Merlin speaks only in the third person, which lends the column its all-knowing "Miss Manners" style.

Over my years of writing the column I have benefited from comments and corrections through consultation with many colleagues, but especially from Drs. Jeff Brown, the late Harlan Smith, Tom Barnes, Rick Binzel, Mike Rich, and Robert Lupton. As always, my sister, Lynn Tyson, has served as a sounding board and as an anchor to reality for many of my literary experiments before they are penned.

CONTENTS

P_{REFACE}

$\mathcal{J}$ust *Visiting This Planet* is the second collection of real questions about the universe asked by real people and answered by the fictional character Merlin, who is a scholar visiting Earth from the planet Omniscia in the Andromeda galaxy.

Nearly all questions and answers contained in *Just Visiting This Planet* first appeared in the popular-level Q&A column "Merlin" of *Star Date* magazine, published by the McDonald Observatory of the University of Texas at Austin. The questions generally come from a broad mix of working and retired adults, with occasional entries from schoolchildren and even prison inmates.

As a literary vehicle, I have rebuilt the famed Merlin character of Arthurian legend into an educational tool that allows me to explore creative ways of bringing complex topics of the universe within reach of the lay reader.

The units of measure used herein were selected on the basis of whatever was natural for the content and style of the question-answer pair. To decree that all units be on the same system (either scientific, engineering, international, or American) would needlessly muddle much of the content. For example, you wouldn't ask for 1.38-inch film for your camera when it's labeled 35mm. And we all know to ask for a half-gallon of milk at the supermarket, but the same supermarket also sells cola in two-liter bottles. In the end, I valued frictionless communication over intemperate rules.

The Merlin character simultaneously embodies my enthusiasm for astrophysics and my daily desire to share cosmic discovery with the public. Your enjoyment will be my reward.

Neil de Grasse Tyson
New York City
January 1998

$\mathcal{M}$erlin is as old as Earth and has been an eyewitness to nearly all major human discoveries in the physical sciences. On temporary leave from home—the planet Omniscia in the Draziw star system of the Andromeda galaxy—Merlin has chosen Earth and its scientific legacy as a topic of academic research. Merlin has advanced degrees in astrophysics, geophysics, chemistry, and philosophy, all earned at Omniscia's planetwide Universe-ity.

A consummate scholar-educator, Merlin loves nothing more than to answer your questions about our favorite universe. Steadfastly succinct yet always friendly, Merlin's replies are occasionally enriched by recollections from the past—a past that contains a motley assortment of Merlin's Earth-based friends and acquaintances, including Archimedes, Galileo, Einstein, and Santa.

I

YOUR
HOME ZONE

Dear Merlin,

Does the Coriolis effect cause water to swirl down drains in different directions in different hemispheres? What about on the equator?

BILL DELLINGES

NEWARK, CALIFORNIA

*I*t depends on the size of your kitchen. If your kitchen sink is a few hundred miles in diameter, then the Coriolis forces induced by the rotating Earth will easily overcome the random water currents and drive the sink to empty its contents in a counterclockwise swirl. If you have Southern Hemisphere friends with an equally large kitchen sink, then theirs will indeed empty in the opposite (clockwise) direction.

When sinks and swimming pools and tubs and toilet bowls are of conventional size, they are not large enough for Earth's Coriolis forces to dominate. This is why, contrary to some legends, you cannot use the "toilet bowl test" when you are dropped in an unknown part of Earth and you want to know what hemisphere you fell in.

On the equator, where there is no Coriolis force, people do not meditate upon their plumbing fixtures.

Dear Merlin,

How come the North Pole is warmer than the South Pole?

HANNAH YOUNG

RICHLAND, WASHINGTON

There are two dominant reasons:

1. There is no land near the North Pole—it is just a spot on a floating ice sheet in the Arctic Ocean. Heat stored in the ocean over the summer months is slowly reradiated to the ice and air during the winter months. In contrast, the South Pole is centrally located in the Antarctic continent, far away from the Antarctic Ocean.
2. Earth is also about 3 percent farther from the Sun during the Antarctic winter (July) than during the Antarctic summer (January).

When combined, these two effects provide a midwinter South Pole temperature of about −100 degrees Fahrenheit, whereas a typical midwinter North Pole temperature is a toasty −40 degrees. This is why Merlin recommended the North Pole over the South Pole when Santa was looking for a home.

Dear Merlin,

Where on the surface of planet Earth is the force of true and apparent gravity equal? Is it the North Pole, the South Pole, or the equator? Could it be 30 degrees south of the equator or 30 degrees north of the equator?

DAVID E. HENDRICK

LEMON GROVE, CALIFORNIA

The only places where you weigh what Earth's mass wants you to weigh is at the North and South Poles. Yes, Santa is the only person on Earth who routinely knows his exact gravitational weight.

At any other latitude, from the poles to the equator, the centrifugal forces of Earth's rotation subtract weight from you. But before you run to the equator to weigh yourself, consider that most household scales lack the precision to measure the several-ounce weight loss expected for a 150-pound person.

Dear Merlin,

What would happen if I dropped a golf-ball-size black hole from waist level onto the floor? Would it go through Earth, pop out the other side, and then fall back repeatedly, turning Earth into Swiss cheese?

BILL DELLINGES

NEWARK, CALIFORNIA

*I*n an encounter between Earth and a golf-ball-size black hole, Earth would lose.

A golf-ball-size black hole has over three times the total mass of Earth. When you release the black hole (from the waist-level pocket of your black-hole-proof suit), Earth and the black hole will fall toward each other until the black hole consumes the entire Earth. Earth will be systematically crushed and funneled across the black hole's event horizon. As you watch this happen, you may regret the experiment.

For having eaten Earth, the black hole will grow to be about 30 percent wider—about the size of a lime.

Dear Merlin,

> *At what altitude does Earth's atmosphere end?*
> Jason Williams
> St. Paul, Minnesota

*E*arth's atmosphere ends where you can't tell the difference between the air you breathe and the space you breathe— where the atmosphere's gas density equals the very low gas density of interplanetary space. Normally, this blend can be found several thousand miles above Earth's surface.

During peak solar activity, however, Earth's atmosphere receives extra heat and expands. The boundary layer can grow outward up to an extra thousand miles and threaten to increase the air resistance on the otherwise stable orbits of high-altitude satellites.

Dear Merlin,

    What is a geosynchronous orbit? What is the difference between a geosynchronous and a geostationary orbit? Are they the same thing?

    MELANIE FRIEDLANDER

    JACKSONVILLE, FLORIDA

*H*ave you ever seen an anchored satellite dish in somebody's back yard and wondered how it could point to something that was in orbit around Earth? The answer contains a clever triumph of human understanding of gravity.

Orbits near Earth's surface complete one revolution in about 90 minutes. Orbits that are higher up will always take longer; consider that at 240,000 miles, the Moon's orbit takes nearly a month. At 22,237 miles above Earth's surface, an orbit takes 23 hours, 56 minutes, and 4.1 seconds—just the time it takes Earth to complete one rotation. All satellites in this orbit, when they move west to east, are geosynchronous because they revolve with Earth. For communication satellites it is the orbit of choice, and, of course, it is what enables broadcasters occasionally to claim *"Live, via satellite."* What the broadcasters don't tell you, however, is that the round-trip signal (Earth-satellite-Earth), at the speed of light, incurs no less than a one-quarter-second time delay. They should really say *"Almost live, via satellite."*

A typical geosynchronous orbit is tipped relative to Earth's equator. In the sky, a satellite with this orbit will never rise or set but will oscillate north-south each day. Geosynchronous orbits that are not tipped relative to Earth's equator will not oscillate. They are truly geostationary. A geostationary orbit is therefore a special case of a geosynchronous orbit.

Dear Merlin,

Years ago a friend and I were bored after dinner. So we calculated that the kinetic energy of Earth moving around the Sun is actually greater than the potential energy of a gas tank the size of Earth filled with gasoline. That is, a gallon-size rock in the yard has more energy than a gallon of gas because the rock is swirling around the Sun at 65,000-odd miles per hour. Would this kind of calculation be considered within the field of astronomy? If so, were our calculations correct?

R$_{ICK}$ R$_{AYFIELD}$

F$_{AYSTON}$, V$_{ERMONT}$

You must have been very bored indeed. Merlin hopes that you did not suffer after-dinner indigestion from your calculations.

Merlin's abacus verifies that you are correct. The kinetic energy that Earth has bestowed upon a gallon-size rock is about a hundred times greater than the chemical-potential energy of a gallon of gas (high-octane, "regular," or "economy").

P.S. Astronomers do not normally think about converting Earth into gasoline.

Dear Merlin,

We are told in school that Earth rotates at a specific rate. Why can't I feel it?

TONY F. GAIK

SEDONA, ARIZONA

If you have ever eaten at one of those fancy large rooftop restaurants that rotate slowly, you will recall that it is hard to notice the rotation unless you look out the window to see the passing scenery. Slow and steady rotation is always difficult to feel.

Earth is larger and takes longer to complete a full rotation than fancy rooftop restaurants. For Earth, of course, the passing scenery is the Sun, Moon, and planets in the sky.

Dear Merlin,

    Please tell me the following speeds: How fast does Earth spin? How fast does Earth go in its orbit? How fast does the Sun move around our galaxy? How fast is our galaxy moving?

    JACK LARNED

    SAN ANTONIO, TEXAS

A spot on Earth's equator travels nearly 25,000 miles (Earth's circumference) in a day—a speed just over 1,000 miles per hour.

    The planet Earth moves over 580 million miles in a year around the Sun—a speed of about 18 miles per second.

    The entire solar system moves almost 1 quintillion miles around the galactic center in 24 million years—a speed of about 125 miles per second.

    The entire Milky Way galaxy (along with the Andromeda galaxy) falls at a leisurely 200 miles per second toward the great Virgo cluster of galaxies.

Dear Merlin,

Do westerly trade winds blowing over north-south mountain ranges keep Earth rotating?

L. D. GRAVES
SUISUN CITY, CALIFORNIA

No.

Earth does not need a force to keep it rotating. On the contrary, Earth would need a force to stop it from rotating. In the late seventeenth century, Merlin's good friend Isaac Newton told Merlin, "Every body continues in its state of rest or of uniform motion in a right line unless it is compelled to change that state by forces impressed upon it." In twentieth-century English, Newton meant that an object in motion (or at rest) will remain in motion (or at rest) unless acted upon by an outside force. His law was written for objects moving in straight lines, yet it also holds for objects that rotate uniformly.

Regardless of all this, the altitude range of the westerly trade winds is above 23,000 feet, which is higher than all of Earth's north-south mountain ranges.

Dear Merlin,

Does the Moon rotate? Does the Sun rotate? If so, what started it?

LYNN MCKERNON

EL PASO, TEXAS

Yes. Yes. Nothing.

Objects in the universe are born rotating. When gas clouds collapse to form planets, stars, and galaxies, they actually rotate faster, following exactly the same law of physics that permits platform divers to spin faster when they tuck during a dive—a principle generally called the conservation of angular momentum. In the universe, as Merlin's good friend Galileo Galilei's experiments first showed in the sixteenth century, there is no reason why an object in motion (this includes rocking and rolling and rotating) should ever stop unless friction or some other force tries to stop it.

Rotation is such a natural feature that the question "Is there anything that does not rotate?" is actually more appropriate. A nonrotating object is a very special, rare case, which in fact has never been observed.

Dear Merlin,

If a basketball court were laid out in an east-west direction, and if I were to shoot a basketball east to west from a distance of forty-five feet to the basket (that is, in a direction opposite to the rotation of Earth), and postulating that the basketball would be airborne for about two seconds, would one have to adjust the basketball shot to take into account the rotation of Earth in the direction from which the ball has been shot?

LAWRENCE R. ROSS, M.D.

NEW YORK CITY

If you have ever jumped in the aisle of an airborne airplane, you may have noticed that you did not fly backward down the aisle and become pinned to the last seat next to the bathrooms. This is because you and the plane were traveling at exactly the same speed—before, during, and after your jump.

Similarly, we find that at 41 degrees north (the latitude of your home town), the New York Knicks basketball team, the arena in which they play, the ball they dribble, shoot, and slam dunk, the cheering fans, the air they breathe, and every other person, place, or thing at 41 degrees around the world moves east with the rotating surface of Earth at exactly the same speed (about 680 miles per hour). An Earth-corrected basketball shot is therefore unnecessary.

Dear Merlin,

With all the motion of Earth through space, when I jump, how come I don't land in a different place?

JOSHUA WALTON

DUVALL, WASHINGTON

You do land in a different place in space, but the ground stays with you.

If you jump straight up and manage to stay airborne for one second, then you will land over 500 yards east (the rotation of Earth carried you there) and 18 miles farther around the Sun (the orbit of Earth carried you there) and about 125 miles farther around the center of the Milky Way galaxy (the orbit of the solar system took you there).

Incidentally, all this happens in one second even if you do not jump.

Dear Merlin,

Why is Earth tilted at 23½? Also, why are the other planets tilted?

JOHN BERKLICH

HIBBING, MINNESOTA

*O*ver the 41,000-year precession-nutation cycle, where it wobbles and bobs like a top, Earth's tilt varies from about 22 degrees up to about 24½ degrees. It just happens to be 23½ degrees at the moment.

The tilts of other planets are related to the detailed conditions that prevailed when the original jumbo gas cloud collapsed to form the solar system. Some planets—most notably Uranus, with a tilt of 98 degrees, and Pluto, with a tilt of 122 degrees—may have suffered major planet-tipping impacts by leftover chunks of debris during the late formation stages of the solar system.

Dear Merlin,

Would you please explain why the Moon doesn't rise consistently fifty minutes later every night of the year?

LAURIE AND STEVE FISHER
CLARKSVILLE, ARKANSAS

The rising time of the Moon has no special affinity for whole numbers that are divisible by ten.

The Moon's speed in its elliptical orbit varies throughout the lunar month, and the position of the Moon along with its rising angle to the horizon vary throughout the year. These factors combine to induce a small daily change in how much later the Moon will rise. On average, the Moon rises 48 minutes and 46 seconds later each day.

Dear Merlin,
    What is a blue moon?
      LINDA HAYDEN, MOONMAIDEN
      ARCADIA, SOUTH CAROLINA

*I*n the absence of blue-tinted moon glasses or peculiar atmospheric optics, a blue moon is simply the second full moon in a calendar month. With months of twenty-eight, twenty-nine, thirty, and thirty-one days and the average cycle of full moons requiring twenty-nine and a half days, the average time between blue moons is about two and a half years.

Occasionally, however, blue moon enthusiasts get a treat. In 1999, 2018, and again in 2037, January has two full moons. In each year, February follows with no full moon, but then March dons two full moons of its own.

Dear Merlin,
  Can you please explain the Harvest Moon?
  LAURIE AND STEVE FISHER
  CLARKSVILLE, ARKANSAS

The nearest full moon to the first day of autumn (September 21–23) used to be called the Harvest Moon since it permitted farmers to work past sunset and complete their last harvest before winter by the light of the Moon.

"Harvest Moon" is also, of course, one of Merlin's favorite jazz compositions.

Dear Merlin,

The Moon has many craters and holes, and they all seem to have occurred long ago. Do they still happen?

*K*RISTEN *O'K*RAGLY

*H*OUSTON*, T*EXAS

*I*ndeed, most of the Moon's craters were formed long ago, during the first billion years of the solar system. Back then, interplanetary space was a hazardous place with countless leftover planetessimals, comets, and asteroids.

But solar system jetsam remains, and it continues to pose a hazard to the Moon, Earth, and all other planets and their satellites. If you do not believe Merlin, just ask Jupiter. It got slammed by the two dozen chunks of comet Shoemaker-Levy 9 in July 1994, leaving Jupiter's atmosphere temporarily scarred.

Dear Merlin,
   Why does Earth have craters?
   LAURA JUDD (AGE 10)
   PHOENIX, ARIZONA

Compared with the Moon, Earth barely has any craters. But you live in Arizona, so you probably have a rather different view of Earth's surface—your home state is world-famous for its holes in the ground. Among these is the Barringer Crater. Since it has a diameter of nearly fourteen football fields, a single visit can leave a big impact.

Meteor craters are formed by meteors that strike the surface. Earth's atmosphere protects the surface from most of the smaller meteors by vaporizing them in a streak of light called a shooting star. Occasionally, however, a big one gets through and simply leaves a big hole in the ground.

Dear Merlin,

   Lately I have been hearing that people may soon go to the Moon to establish colonies for mining. If this does happen and large amounts of Moon matter are brought to Earth, wouldn't the ratio of gravitational attraction between the Moon and Earth change? Isn't there a chance that the orbit or speed of the Moon would change? If so, would we have a change in the tides? Would the ocean become stagnant, as the plant and animal life regulated by the tides would just sit there?

   Seems to me that the gravity of the Moon would lessen and the gravity on Earth would heighten and we would all weigh more. Wouldn't that cause health problems? Please check this out, Merlin. I am worried.

   PHYLLIS A. LINN

   LOMITA, CALIFORNIA

*I*f the Moon were mined for its resources and if all this material were brought back to Earth, then yes, gravity on the Moon would drop, gravity on Earth would increase, the tidal forces of the Moon on Earth would lessen, and the Moon's orbiting distance would increase.

   But let's imagine an extreme case: suppose the entire Moon were hauled back to Earth piece by piece. Earth is so much more massive than the Moon that when the task was completed, a person who normally weighed 150 pounds would weigh about 150 pounds, 10 ounces—hardly an occasion to alert the World Health Organization.

Dear Merlin,

I am intrigued when I look at the Moon when it is not full and I see a faint outline of the darkened portion. Is this a part of the lunar hemisphere that we never see (the dark side of the Moon), or is it the same lunar hemisphere always facing us with the position of the lunar dawn simply moving across it?

JIM TODD

SAN DIEGO, CALIFORNIA

Contrary to popular musical literature, folklore, and best-selling rock albums, there is no "dark side" of the Moon. Indeed, the position of the lunar dawn migrates across the entire lunar surface to provide nearly fifteen consecutive days of sunlight to every part of the Moon. In scientific parlance, this moving boundary between light and dark has the less than poetic name of "terminator."

The Moon does, however, display only one side—the "near side"—to Earth at all times. In late 1959, the Soviet spacecraft *Luna 3* flew past the Moon. Only then did earthlings obtain the first photographs of the Moon's "back side."

Dear Merlin,

It always fascinates me that we see only one side of the Moon. Why is its rotation so in sync with Earth's? Surely it is not just a coincidence of nature, is it?

MIKE NOLIN

BREWSTER, MASSACHUSETTS

*B*illions of years ago, the Moon rotated on its axis much faster than once per month, but tidal forces from Earth on the Moon served to slow its rotation rate until the Moon rotated exactly once for every period of revolution around Earth. This arrangement is known as a tidal lock, and it manages to hide the far side of the Moon from Earth view forever.

At the moment, Earth rotates faster on its axis than once per month. The tidal effects of the Moon on Earth are slowing down Earth's rotation so that eventually the Earth day will equal the lunar month, which is known as a double tidal lock.

These terms, which sound like advanced wrestling holds, refer to common configurations in the universe. For example, Jupiter's moon Io is in tidal lock with Jupiter, and Pluto has achieved a double tidal lock with its lone moon, Charon.

So, contrary to suspicion, synchronized rotations are not the result of cosmic coincidences.

Dear Merlin,

In the 1983 movie Superman, *there is a scene where Superman flies around Earth in the opposite direction to its rotation. In the process, he manages to reverse Earth's rotation and thereby reverse time. Is it possible to reverse time in this manner, or was that just Hollywood being Hollywood?*

Arnold Weaver
Los Angeles, California

You seem to have no trouble believing that a man with superhuman powers and X-ray vision can fly through the air wearing panty hose and a cape. Merlin sees no reason why you shouldn't believe everything else in the movie.

If indeed Superman managed to fly west around Earth and reverse Earth's rotation, the flow of time would remain unaltered. But everything that was not bolted to Earth (humans included) would fall over and roll due east at hundreds of miles per hour.

Dear Merlin,

If Earth's rotation is gradually slowing down, and as a result the Moon is receding from us, does this mean that if we could somehow speed up Earth's rotation with giant rockets, the Moon would spiral in on us?

KARL RIGHTER

ALTAMONTE SPRINGS, FLORIDA

The tidal forces from the Moon cause the Earth's rotation to slow (which indeed causes the Moon to spiral away at a rate of about an inch per year), but you have mixed your causes and effects. The Moon is not accountable for how you play with your rockets.

If you managed to speed up Earth's rotation, all that would happen is that the days would become shorter, the precession period of the equinoxes would become longer, and everybody on Earth (except Santa Claus) would get jolted to the west while the engines fire.

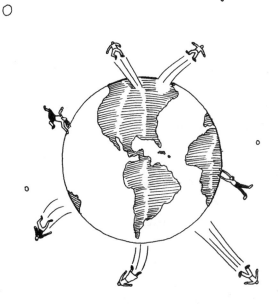

Dear Merlin,

If aliens exploded our Moon, what effect would it have on us?
LARRY BOSWELL
AUSTIN, TEXAS

If angry aliens exploded Earth's moon, then most astronomers, believe it or not, would be quite happy. The full moon, and phases close to it, illuminate the night sky brightly enough to render "deep sky" astronomy impossible. In other words, observing the distant universe when the Moon is full is like watching a drive-in movie in the daytime.

An exploded Moon would otherwise not make much difference to life on Earth. Tides would not rise as high (they are still affected by the Sun), eclipse chasers would lead boring lives, and people would have to find something else upon which to blame "lunatic" behavior.

Dear Merlin,

I know the National Enquirer *is often incorrect, but recently it ran a very convincing article saying that we never really got to the Moon at all— it was all a hoax. Seems like the government would have challenged the paper if this claim were untrue. Please tell me, did we get to the Moon or not?*

LORRAINE MONTE

LADY LAKE, FLORIDA

*W*as the Moon-hoax article more convincing than:

1. the thunderous blastoff of the Saturn V rocket that contained the *Apollo* astronauts?
2. the two-and-a-half-second time delays during the live phone conversation that President Nixon, on Earth, conducted with Neil Armstrong, on the Moon? (It takes a phone signal two and a half seconds to make the 480,000-mile round trip.)
3. the films of twelve astronauts (two each from *Apollo* 11, 12, 14, 15, 16, and 17) bouncing about on the barren, dusty lunar surface?
4. the countless photographs of Earth as seen from the Moon?
5. the Moon rocks that were brought back to Earth and donated to every major museum in the world?
6. the detected laser signals, sent from Earth to the Moon, that reflect from specially angled mirrors left by the astronauts?
7. the splashdown of the command modules and their charred undersurface from the heat of reentering Earth's atmosphere?

Yes, humans have been to the Moon, whether or not a newspaper can sell more copies based on claims to the contrary.

Dear Merlin,

If people were living on the Moon, would they see phases of Earth as we on Earth see phases of the Moon?

CAROL GARCIA

KILGORE, TEXAS

A patient observer placed on the near side of the Moon for a month will see "new earth" during full moon, "full earth" during new moon, and all phases in between.

Our patient observer will also notice that Earth never sets, and that "full earth" seen from the Moon is over fifty times brighter than full moon seen from Earth.

II

PLANETS,
THEIR MOONS,
and OTHER
COSMIC DEBRIS

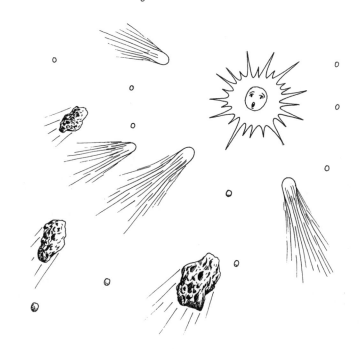

Dear Merlin,
   Could you list all the planets in order from the Sun for me?
   V. SCHWARTZ
   TRENTON, NEW JERSEY

$\mathcal{M}$erlin always remembers the planets in order of the distance from the Sun with the following mnemonic: "My very educated mother just served us nine pizzas." The first letters match the first letters of the planets in order: Mercury, Venus, Earth, Mars, Jupiter, Saturn, Uranus, Neptune, Pluto.

But for twenty years in its 248-year elongated orbit (as was true between 1979 and 1999), Pluto is closer to the Sun than Neptune is. During those years our mnemonic should read: "My very educated mother just served us *pizzas nine.*"

Dear Merlin,

Ancient peoples, without telescopes, would have no knowledge that Jupiter is the largest and most majestic planet—it would only be a point of light in the sky. How then did this planet come to be named for the mythological ruler of the gods? For that matter, how did the planets come to get the names they now have?

WILLIAM W. BURNS

BARTLETT, TENNESSEE

The innermost planet in the solar system is not only the fastest in its orbit, it is the fastest when viewed against the background sky. The Romans named this planet after Mercury, their fleet-footed messenger god. Incidentally, those little winglets on Mercury's shoes and hat are aerodynamically useless.

The second planet from the Sun is by far the brightest and most beautiful in the twilight sky. The Romans aptly named this planet after Venus, their goddess of love and beauty and all that goes with it. But little did they know that a runaway greenhouse effect under Venus's thick cloud cover has made its surface temperature high enough to melt zinc and vaporize people. And little did they know that sexually transmitted diseases would one day be named "venereal" in Venus's honor.

The Romans were conquerors, and thus were skilled at killing people. When this sort of thing excites you, then you naturally associate the color red with bloodshed rather than with roses. It was an easy choice: the lone red planet in the sky was named for Mars, their god of war.

The slowest and most distant planet known to the ancients was traditionally identified with age and wisdom. (Merlin has actually never met anybody who was old and wise yet in a hurry.) In Roman mythology, this planet was named for the elder Saturn, one of the race of Titans.

The one planet that moved faster than Saturn but slower than all the others was named Jupiter—an obvious choice when you realize that Jupiter is the son of Saturn and the father of Mercury, Venus, and Mars.

The fact that the planet Jupiter, named for the ruling Roman god, is the most massive planet is simply a coincidence. Had Saturn been the most massive planet, one could argue just as convincingly that Jupiter's Titan father was deserving of the largest mass.

The "modern" planets Uranus, Neptune, and Pluto were each named by committee to keep them part of one big happy mythological family: Uranus is Saturn's father, and Neptune and Pluto are Jupiter's brothers.

The Sun, Earth, and Moon are not named after gods. Each name is derived from an Old English form.

Dear Merlin,

Are all of the planets in the solar system in the same plane? If so, does the plane coincide with the equator of the Sun?

WILLIAM R. DONOP

MASON, TEXAS

If you neglect Pluto's orbit, which is inclined 17 degrees from the plane of Earth's orbit, all the planet planes (as well as the equator of the Sun) lie within about 7 degrees of each other.

This fact, combined with the knowledge that all planets orbit the Sun in the same direction, led Merlin's good friend the astronomer-mathematician Pierre Simon Laplace to suggest a "nebular hypothesis" for the formation of the planets. In his 1796 *Exposition du système du monde*, he held that the solar system started from a rotating gas cloud that flattened and collapsed into zones, from which the various planets were formed—a thought advanced somewhat earlier by the German philosopher Immanuel Kant.

Comets are legitimate members of the solar system but can orbit the Sun at all angles, even perpendicular to the flattened plane. These interplanetary vagabonds of prehistory were too distant to participate in the collapse and flattening of the original gas cloud.

Dear Merlin,

I finally looked up the term "orrery," which I occasionally see in astronomy articles. Did you know it comes from Charles Boyle, fourth earl of Orrery? I guess Orrery sounds better than Boyley!

ARTHUR WOODS
COPPERAS COVE, TEXAS

*N*ot only did Merlin know this tidbit, but Merlin was there when Charles Boyle's first mechanical model of the solar system was built by John Rowles in the early eighteenth century. Rowles had a large collection of mechanical paraphernalia with which he liked to make friends and influence people. His first orrery is on display at the Science Museum in London.

Incidentally, Merlin agrees that Orrery sounds better than Boyley, but they both look like typographical errors.

Dear Merlin,

In an encyclopedia from 1885 I found a reference to a planet inside the orbit of Mercury called Vulcan. Do you know what this was referring to?

MIKE STILES

SAN LUIS OBISPO, CALIFORNIA

It was referring to the planet Vulcan, inside the orbit of Mercury.

No such planet existed then, or now. When scientists discovered that Mercury's orbit did not precisely follow the predictions drawn from Isaac Newton's laws of gravity (after all known sources of gravity were accounted for), a hypothetical planet, Vulcan, was invented. Its gravity would enable Mercury's orbit to obey Newton's laws.

In 1916, Albert Einstein published the general theory of relativity, which redefined human understanding of gravity. Einstein's theory removed the need for any hypothetical planetary perturbers. Vulcan has since joined the science graveyard, along with ether, caloric, the steady-state theory of the universe, and other ideas that have come and gone.

Dear Merlin,

   Why don't plants grow on Mars? Is it because the soil isn't good, or is it because there is no water?
   R OBYN A BRAMS
   C LONIS , C ALIFORNIA

Ordinary Earth plants require sunlight, soil nutrients, carbon dioxide, oxygen, water, and atmospheric-blanket protection from the Sun's high-energy ultraviolet light.

   Mars lacks oxygen and an effective ultraviolet shield. The Martian atmosphere is mostly carbon dioxide and is very thin compared with that of Earth. The high-energy ultraviolet light that reaches the Martian surface would decompose life as we know it.

   Flowing water existed long ago on the surface of Mars—there are meandering dry riverbeds, river deltas, and river basins—but its water is now locked in polar ice caps and subsurface permafrost.

   That being said, we cannot rule out the possibility of simple anaerobic life deep underground.

Dear Merlin,

I have been reading recently about the proposed trips to Mars. In many scenarios there seems to be a roundabout approach to getting there. One scenario even uses the gravity of Venus as a sort of slingshot to Mars.

What I don't understand is why we don't take a direct approach like skeet shooting. That is, launch our ship in an Earth orbit until the time is right, then leave Earth orbit when the orbit of Mars is closest so that Mars will move into the spot of interception.

LESTER HELLMAN

PRESCOTT VALLEY, ARIZONA

*I*nterplanetary travel can be a tricky business. In a civilization where public policy is guided by economics and the availability of resources, the easiest path between two points is not always the shortest distance.

To conserve fuel, those in charge often choose a path through space that flies past a planet along the way, because another planet's gravity costs less than rocket fuel. This is the famous gravitational slingshot effect, which has been used for almost all planetary probes. The resulting path may be twice as long as the shortest distance, but the ride is practically free.

Dear Merlin,

I recently read an interesting article on Mars and its early history. The theme was that life could have developed on Mars during its early history and that remnants or perhaps fossils could be awaiting either robot or human explorers. What are your ideas on what type of life may have developed?

ROY KRAUSE
CLEAR AFB, ALASKA

If the history of life on Earth is any indicator, Mars may have once teemed with simple, single-celled life forms and possibly primitive plants clustered near the flowing rivers. A life-hostile change in Martian climate may have taken place early enough to arrest the evolution of complex life and, of course, the emergence of LGM,* which means the riverbeds may hold the fossil remains of an extinct biota.

* little green men

Dear Merlin,

A neighbor claims (from watching a television program) that there are discernible faces and pyramids apparently manufactured on the surface of Mars. His opinion is that if this were general knowledge, the people on Earth would panic. Can you verify or refute this information?

JACK D. CULP

MORONGO VALLEY, CALIFORNIA

Shadows are notorious for what they can "apparently manufacture" in the mind.

The images of the Martian surface beamed back to Earth by the *Viking* missions in the mid-1970s showed a shadowy, rocky terrain. The news media were drawn to several patterns that vaguely resembled masks, numerals, pyramids, and other shapes that shadows mimic easily.

What went unnoticed were 12,932 other shadowy patterns that did not look like anything but rock shadows.

Puffy cumulus clouds can play the same tricks. Surely you have seen faces in cloud shapes. Did you panic?

Dear Merlin,

I've heard so many interesting things about the planets in our solar system, but one thing still puzzles me. I know Jupiter is mainly gas with a red eye, but is Jupiter hot or cold?

RENEE TINSLEY

STOCKTON, CALIFORNIA

The latest planetary fly-bys confirmed what has been suspected for some time: Jupiter does not have eyeballs. Jupiter does, however, have quite a few regions that are officially called spots. The big red one is called Jupiter's Big Red Spot.

The temperature in Jupiter ranges from a chilly −150 degrees Fahrenheit near its gaseous surface to a toasty 35,000 degrees Fahrenheit deep in its liquefied core. Remember that Jupiter was almost a star when it formed, and it still contains a substantial amount of this original heat. Consequently, Jupiter radiates more energy into space than it receives from the Sun.

Dear Merlin,

We take it for granted that we can see through our own atmosphere, yet there are other planets, such as Jupiter, that we cannot visually penetrate. Would it be possible that if life did exist on one of these visually opaque worlds, those life forms could see through their atmosphere but not ours?

CHARLES BURKE

FARMINGDALE, NEW JERSEY

No.

Dear Merlin,

Couldn't the fragments of comet Shoemaker-Levy 9, when they exploded on the surface of Jupiter back in 1994, cause the massive planet to move? Would any movement throw Jupiter's moons off course?

JENNIFER SIMPSON

MIDLAND, TEXAS

A n elephant is a massive animal. Jupiter is a massive planet. A gnat that flew full speed ahead and slammed broadside into an elephant would be about one million times more effective at shoving the elephant than comet Shoemaker-Levy 9 was at shoving Jupiter.

Jupiter, which is more massive than all the other planets combined, would need much more than a pesky comet encounter to measurably alter its own orbit or that of its moons.

Dear Merlin,

When the spacecraft Galileo *dropped its probe into Jupiter's atmosphere in 1995, was it the first time spacecraft fragments dropped anywhere but on Earth?*

BESS WHITMORE

BRONSON, FLORIDA

The remains of human-made hardware can now be found strewn on the Moon, Venus, Mars, Jupiter, across interplanetary space, and vaporized in the Sun. The sundry mixture contains Soviet and American flags, cameras, probes, antennas, electric automobiles, golf balls, and dozens of lost and crashed satellites.

Dear Merlin,

In his Adventures of Captain Bonneville, Washington Irving reports that in 1832, "by an observation of Jupiter's Satellites . . . Captain Bonneville ascertained the longitude to be 102 degrees 57 minutes west of Greenwich." (Bonneville was off by a couple of degrees.) I know that Galileo recommended observation of the eclipses of Jupiter's moons to determine longitude, but how does it work?

MARY BOWDEN

AUSTIN, TEXAS

Jupiter's four brightest moons occasionally go in front of and behind each other, or they simply occupy predictable locations in their orbits. If you know the orbits of Jupiter's satellites very, very well, and if the predicted eclipse times are tabulated for a reference time zone, then you can easily deduce your longitude on Earth based on the local time you saw them or what time you noted a particular arrangement through your telescope.

Galileo knew the importance of the effort. Always eager to drum up interest in his "spyglass," Galileo said to Merlin one afternoon in Florence back in 1613, "That great and marvelous problem of finding the longitude of a given place on Earth's surface has been so much desired in all past centuries because of the important relations of such a discovery to the complete perfection of geography and naval maps."

Galileo's newfangled telescope finally found an application that was practical. Other uses had been considered blasphemous and trivial—for instance, confirming that Earth is not the center of the universe.

Dear Merlin,

Can you tell me who or what discovered the planet Saturn?
MASHA GALEB
BATESVILLE, ARKANSAS

No. There are six planets that are easily visible to the unaided eye: Mercury, Venus, Earth, Mars, Jupiter, and Saturn. They are all considered "historical" planets, with no traceable discoverer. In other words, they have always been visible to anybody who has ever looked up (or down), including Trog & Lodyte, early cave friends of Merlin's.

For the record, William Herschel discovered Uranus in 1781, Johann Galle discovered Neptune in 1846 from a prediction by Urbain LeVerrier, Clyde Tombaugh discovered Pluto in 1930, and there is no Planet X.

Dear Merlin,

Why does Neptune's moon Triton orbit backward, in contrast to all other moons in the solar system? Could a large body have disrupted the orbits of the moons of Neptune long ago? Does the latest Voyager data support the possibility that the Neptunian system could have been disturbed by a close encounter with the so-called Nemesis object?

Roy Krause

March AFB, California

*N*emesis, the hypothetical binary star companion to the Sun, has never been observed. If it does exist, and if it did have a close encounter with Neptune, we would expect more damage than just perturbed satellites—Neptune would not likely have preserved its near-circular orbit around the Sun.

Triton's unusual orbit suggests that it may have been an interplanetary interloper that was captured by Neptune in a close encounter sometime after the Neptunian system was formed.

Dear Merlin,

Do the other planets have cardinal directions as we understand them here on Earth (i.e., north, south, etc.)?

Darrell Kruzen
Wisdom, Montana

$\mathcal{D}$oes everyone from your town ask wise questions?

All rotating objects have north and south poles. By convention, if you curl the fingers of your right hand in the direction an object rotates, then your extended thumb will point due north. It works for planets, and it works for the entire Milky Way galaxy. More generally called the right hand rule, it enjoys wide application in branches of physics such as classical mechanics, electrodynamics, and quantum mechanics. And yes, physicists make excellent hitchhikers.

Dear Merlin,

Could we use ordinary field compasses on the other planet surfaces to navigate from point to point? If not, why not?

DARRELL KRUZEN

WISDOM, MONTANA

On Earth, if you hike anywhere in Alaska, northern Canada, or Greenland, a field compass will get you lost. Depending on your exact location, compass north can be north, east, west, south, or any direction in between.

There is nothing cosmic about this. Earth's magnetic north pole happens to be nearly one thousand miles away—due south, of course—from the geographic North Pole. (At Merlin's prompting, Santa Claus no longer uses a compass to return home. He uses a global positioning satellite receiver in the dashboard of his sleigh.) If you walk north as instructed by your trusty field compass, you will wander among the Queen Elizabeth Islands of northern Canada. There is no law of physics that says the two poles must perfectly coincide.

Let us ignore the complication that your compass will melt on the sunlit side of Mercury, and that Jupiter, Saturn, Uranus, and Neptune have no solid surface. On Jupiter the magnetic field is nearly twenty thousand times stronger than on Earth, but your compass will point south, not north. And on Uranus and Neptune you should leave your compass at home—their magnetic poles are tipped closer to their equators than to their "geo"graphic poles.

On Venus, Mars, and Pluto your compass will be of no use. These planets have no measurable magnetic field.

Dear Merlin,

Instead of burning up in Earth's atmosphere, is it possible for a meteor to be captured by Earth's gravity and become a natural satellite?

HANS K. SCHUBERT

LYNCHBURG, VIRGINIA

Yes.

Dear Merlin,

    As space probes discover more moons and asteroids, how are they named and by whom? Also, are there any moons or space objects named Merlin?

    R OY K RAUSE

    M ARCH AFB, C ALIFORNIA

*A*ll planets except Earth are named for gods from Greek and Roman mythology. All moons except Earth's moon (which is named simply Moon) and the satellites of Uranus (which are named after Shakespearean characters) are named for assorted supporting characters in the life of the god for which the planet is named. For example, to reach Hades, the underworld home of the god Pluto, one must cross the river Styx in a ferry guided by Charon. Pluto is the name of the most distant planet, and Charon is the name of its moon.

    The planets Jupiter and Saturn each have more than fifteen moons. When too many moons are discovered too quickly (as happens in satellite fly-bys), they are simply numbered until the recommended name put forth by their discoverers is agreed upon. Greek and Roman gods had complicated social lives, so there should be no shortage of moon names.

    The discoverer of an asteroid can name it after any person, place, or thing. There are thousands of them, including 2598 Merlin.

Dear Merlin,

I have always thought that the large amount of light given off from meteors was the ionization of the air as the kinetic energy of the particles is absorbed by the atmosphere. Recently, however, I read about a fellow who for many years had been taking the spectrum of meteors. If he is taking spectra, he must be expecting to see emission from the elements that compose the meteor. My question is, what portion of the light given off by a meteor is due to the ionization of molecules in the atmosphere and what portion is due to the ionization of the meteor itself as it disintegrates?

MICHAEL R. RICHARDSON

YONKERS, NEW YORK

$\mathcal{M}$ost of the light you see from a streaking meteor (more commonly termed a shooting star) is from the nitrogen and oxygen molecules in Earth's atmosphere. When heated, they emit specific wavelengths of light that are distinct and clear in spectra taken of the meteor trail.

Upon disintegrating, the meteor itself emits characteristic wavelengths of light that indicate its composition (e.g., iron, silicon, and carbon). This emission may make up less than one percent of the total light, but the spectral signature of the meteor's composition is different and unmistakable when compared with that of the atmosphere.

Dear Merlin,

Recently it was disclosed that a near miss occurred when a small asteroid nearly collided with Earth. Has anyone computed the extent of destruction that such an impact would have wreaked on our globe? Is anyone marshaling global resources to try to detect any incoming debris and divert its orbit?

DONALD K. LEONARD

WESTLAKE VILLAGE, CALIFORNIA

*S*everal families of orbiting asteroids cross Earth's orbit. Most notable (and numerous) among them are the Apollo asteroids. They are typically a mile across and are dim enough to be discovered only when they are relatively close to Earth (that is, when it's too late to do anything about it).

Recent estimates suggest that there are over a thousand Earth-crossing asteroids that are larger than about a half-mile in diameter. The near miss (Merlin prefers "near hit") that you speak of was no doubt the small Apollo asteroid 1989FC, a rock the size of about two football fields that came within 430,000 miles of Earth.

Several telescopes in the world are dedicated to the discovery of near-Earth asteroids. A major problem is that there are more small asteroids than big asteroids. Smaller asteroids are harder to discover and are often quite close before they are detected. Additionally, asteroids are normally discovered by their trail on a time-lapsed image as they move in their orbit. An asteroid that is headed straight for Earth from some directions can appear stationary and escape discovery. Earth currently has no defense against a head-on collision with a sizable asteroid.

Near misses are, of course, rarer than distant misses, but a direct hit is extremely rare when compared with a near miss. A "recent" direct hit by a large asteroid made the 25,000-year-old Barringer Crater in Arizona—the remnant of an encounter with a considerably more common 100,000-ton asteroid.

If tiny 1989FC had hit Earth at maximum speed, its equivalent energy of impact would have been nearly 100,000 megatons of TNT. (Reminder: The atom bombs that the United States dropped on Japan in 1945 were each about 0.02 megatons.) It would have been a most devastating event—earthquakes, tidal waves, climactic upheaval, and mass extinction of life forms. Statistical estimates, however, suggest that a 1989FC-size collision occurs only about once or twice every 100 million years. This interval is enormously longer than the history of the human species.

Unless you live as long as Merlin, you needn't worry about standing in the wrong place at the right time, or the right place at the wrong time.

Dear Merlin,

I once heard, or read, that the asteroid Hermes would soon be passing near Earth in its orbit around the Sun. Is this true?

LEANETTE ASHCRAFT

DEDEDO, GUAM

*O*ne of the closest approaches on record for a substantial asteroid was in 1937, when Hermes was discovered. This 440-million-ton asteroid whizzed within 450,000 miles of Earth, about twice the Earth–Moon distance.

Hermes is in a strongly flattened orbit in the plane of the solar system. Its closest approach to the Sun is about 50 million miles, and its farthest approach is about 250 million miles. Remembering that the Earth–Sun distance is 93 million miles, we can expect an exciting future of close encounters between Earth and Hermes—unless the gravity of Mercury, Venus, or Mars perturbs its orbit and keeps us out of harm's way.

The exact position of Hermes in its orbit is not known well enough to predict with precision the dates of its next close encounters.

Dear Merlin,

If an asteroid killed the dinosaurs, what conditions created by the impact led to their extinction?

JOHN OREM

SHALLOWATER, TEXAS

*A*s is now well known, the single meteor that slammed into Earth 65 million years ago and left a two hundred-mile-wide crater near the Yucatan peninsula contributed to the end of the reign of dinosaurs. No, there was no dinosaur business convention in Mexico at the time. What most dinosaur-killing scenarios have in common is a severe and relatively abrupt climate change, for which the dinosaurs were biologically unequipped.

Merlin had a good friend named Thelma Rex (her friends just called her T. Rex). In her last letter to Merlin she wrote, "When the meteor fell, an enormous explosive plume rose up into the stratosphere and spread round the Earth. Precious sunlight is now being blocked, plant life is dying, and the average air temperature is falling."

Thelma died, of course, but her fossilized brethren are preserved in many of the world's natural history museums so that the now dominant mammals can pay their respects.

Dear Merlin,

I once read a report that claimed to notice a flare from Halley's comet. Was the flare due to a collision with a small asteroid? If so, why wouldn't the collision change the orbit period of Halley's comet?

RON GIFFIN

CINCINNATI, OHIO

The commonly quoted period of seventy-six years for Halley's comet to return is only an average. The true period varies by several years, because unlike planets, comets tend to lead traumatic orbital lives. Their orbits are altered almost daily by a host of causes.

The elongated paths of comets that orbit near the plane of the solar system can move through such cosmic collision zones as the asteroid belt between Mars and Jupiter. Comets occasionally have close approaches with Jupiter or Saturn, which can slow them down, speed them up, or simply sling them off in another direction.

A comet can also lose some (or most) of its mass as the frozen gases evaporate near the radiant Sun. Comets thus lay a flotsam trail in their orbit that can extend millions of miles. Some comets have even been observed to split in two, while others have been eaten by the Sun.

Dear Merlin,

 What is the next bright comet we can see, and when can we see it?
ERIC MORGAN
DOWNEY, CALIFORNIA

The time-honored standard is Halley's comet. It was not especially bright during its past visit, but in the year 2062, Halley should return with all the splendor of its earlier trips.

Many comets are discovered every year. About once per decade, however, a comet is discovered that provides a singular spectacle for all. Comet Hale-Bopp in the 1990s was an excellent example. Often these comets have very long orbital periods—thousands or millions of years. We can guarantee many of these to be brighter than Halley's comet, but we must patiently await their discovery.

Dear Merlin,

I understand that most of the water that existed or still exists on the planets arrived via comets. The dirty snowballs were accreted during the formation of the solar system. I would like to know where the water in the snowballs came from and under what conditions it was formed. Is Earth making water in its interior, or is what we see all we have?

MICHAEL J. ELLIS

PT. REYES STATION, CALIFORNIA

The components of water—hydrogen and oxygen—are quite plentiful in the universe. The water molecule, H_2O, forms readily under a broad range of temperatures and pressures.

There is no doubt that comets contain frozen water (as well as frozen carbon dioxide, frozen methane, and frozen ammonia), and they may have brought the oceans to Earth, but early volcanic eruptions were also a rich source of water.

Earth today is geologically quiet compared with three billion years ago. Yet water vapor (as well as a myriad of noxious gases) still emerges from Earth's mantle with every volcanic eruption. This out-gassed water vapor is not manufactured but simply left over from the original formation of Earth.

Dear Merlin,

Is it true that if I leave a container on my roof for a period of time, I can actually collect particles from outer space?

MICHAEL OYE

OXNARD, CALIFORNIA

*E*arth's entire surface intercepts about one thousand tons of interplanetary particles per day. Few are big, some are small, and most are burned in the atmosphere and descend as meteor dust. At this rate, a frying pan on your roof will collect one ounce of space debris in a little over 300 million days. You can probably find better activities to occupy your time.

People who are in the business of collecting meteorites, however, often use the ice sheets of Antarctica to search for them. The collecting area is large, and unlike meteors that fall in the oceans or in forests, a dark meteor on white ice is readily noticed. This method is somewhat more effective than a rooftop saucepan.

III

STARS,
NEAR and FAR

Dear Merlin,

I would like to know how our Sun continues to burn hydrogen for billions of years. How can the hydrogen supply last so long, and why doesn't it all blow up at once with one big explosion like throwing a lighted match in a can with gasoline vapors inside?

STAN KELLOGG
SAN DIEGO, CALIFORNIA

$\mathcal{M}$erlin does not know why you want to explode the Sun, but consider five factoids:

1. Over 90 percent of all the Sun's atoms are hydrogen.
2. The Sun is very big. If the Sun were hollow, more than a million Earths could fit comfortably inside.
3. The Sun converts about 4 billion (4×10^9) pounds of matter into energy in the time it takes to say "Stan Kellogg."
4. The Sun produces most of its energy (the fusion of hydrogen into helium) within the inner 10 percent of its radius.
5. The Sun's core has about 600 septillion (600×10^{24}) pounds of convertible fuel.

The core is the only place within the Sun that is hot enough (about 20 million degrees Fahrenheit) to sustain fusion. The enormous weight of the rest of the Sun does a splendid job of regulating the energy production. Note that if the entire Sun underwent fusion, the explosion would be somewhat more dramatic than that of an ignited gasoline can.

And if you are handy with your calculator, you may notice that factoids 3 and 5 combine to tell us that the Sun can look forward to many happy years of life—billions of them.

Dear Merlin,

If the Sun were to cease burning and we had enough advance notice (several years), would it be feasible to build domed cities for heat and light that would make it possible for us to exist? What problems other than heat and light would we encounter, and what would the temperature be outside the domed city?

D. C. GLUCK

MANASSAS, VIRGINIA

*N*early all sources of energy consumed by humans on Earth are traceable to the Sun. The list includes all fossil fuels, hydroelectric plants, windmills, and of course rodents on a treadmill. If the Sun stopped producing energy, you could stay warm only for as long as your stored fuels lasted. While nuclear power plants do not depend on the Sun, they will last only as long as the supply of fissionable materials on Earth.

When the energy pantry became bare, you would swiftly freeze, as your bubble-cities radiated their leftover heat into the cold of space while they approached 462 degrees below zero Fahrenheit—the background temperature of the universe. That would be bad.

You may need to pitch your tent next to a river of molten lava near an active volcano—geothermal energy is not traceable to the Sun. Better yet, board the next spaceship off the planet.

Dear Merlin,

Is there any evidence that indicates that increased solar activity can be related to an increase in Earth's geological activity, such as volcanoes, earthquakes, plate movement, etc.?

KARL J. KUEN

KIRKWOOD, NEW YORK

*N*o.

Dear Merlin,

I recently read that a ten-Earth-mass blob of matter was observed being expelled from the Sun. Fortunately, it went off into space in a very different direction from that of Earth.

What I want to know is, had that huge mass of material come directly toward Earth, could it have made a direct hit on this planet like a custard pie in someone's face?

DOUGLAS COLTON

ANGUIN, CALIFORNIA

*B*lobs of matter, which are mostly in the form of hydrogen plasma plumes called solar flares and prominences, are ejected by the Sun all the time.

These eruptions are fast and quite spectacular. Some of the fastest-moving protons and electrons leave the Sun and enhance the ever-present "solar wind." Occasionally a wave of these particles heads directly toward Earth and hits a few days after the eruption. Their arrival can short-circuit the electronics of orbiting satellites, knock out power grids, and trigger the beautiful northern and southern auroras. But in general, the solar gravity has a firm grip on itself. Most of the blobs and plumes return to the Sun's surface without risk of becoming an interplanetary plasma pie.

Dear Merlin,

Our planet is becoming overburdened with atomic waste and hazardous materials at an alarming rate. Why can't we put some of it on space flights when room is available and have the astronauts nudge the stuff in the direction of the Sun?

LESTER DUBACH HELLMAN

PRESCOTT VALLEY, ARIZONA

*W*hat you call a "nudge" from an already orbiting space station amounts to shoving the atomic waste in the right direction at a speed of about 7,000 miles per hour. Only then will it escape the gravitational pull of Earth. If the high-speed nudge fails, the waste will reenter Earth's atmosphere and create an even greater health risk.

The solution for your planet is to find sources of energy that do not produce atomic and otherwise hazardous wastes.

What is solar retrograde motion? I heard about it recently, but I'm not sure I understand it.

STEVE YOUNG
NEW YORK CITY

*E*very object in the solar system orbits the center of mass of the solar system. If the solar system, Sun included, were embedded in a cosmic platter, the center of mass would be the point under which you would have to place your finger to balance it all.

The center of mass is relatively close to the center of the Sun. Most of the time it can be found deep within the Sun's surface. Normally, the planets and the Sun trace systematic loops around this center of mass. But the planets all have different masses, different distances, and different orbital speeds. About every 180 years we find that the arrangement of planets is such that one of the loops the Sun makes does not enclose the center of mass. As seen from the solar system's center of mass, for about twenty years the Sun appears to move in the opposite direction—retrograde—against the background stars.

Some theories suggest that this retrograde period will excite solar activity, thus increasing sunspot counts above the expected numbers. This remains to be demonstrated.

Dear Merlin,

What effect, if any, would a binary star system have on planet Earth?

ALBERT ADELIZZI

CHICAGO, ILLINOIS

*M*erlin presumes you refer to the Sun in a binary system.

If the Sun and a twin companion orbited close to each other in the plane of the solar system then

1. The daytime sky would be twice as bright.
2. There would be frequent star-star eclipses.
3. There would be awesome double sunsets.

If the Sun had a companion more massive than itself, then long-lived earthlings might witness the following spectacle of binary mass transfer:

1. The Sun's massive companion would swell to become a red giant as the tenuous giant envelope became unbound and spiraled toward the Sun.
2. The increased mass of the Sun would induce the Sun to revolve faster and swell to become a red giant too.
3. The mass transfer would then reverse to begin a cosmic egg-toss.

Regardless of the star masses, if they orbited each other at a distance that is comparable to the current Earth–Sun distance, then

1. Earth would not sustain the cozy 93-million-mile separation from the Sun that it now enjoys in its near-circular orbit.
2. There would be tremendous temperature extremes on Earth's surface—Earth's orbit would trace a complicated shape in response to the two dominant and widely separated sources of gravity.
3. Earth's orbit may become so unstable that Earth would either be eaten by one of the stars or be flung off into the void of interstellar space.

Binary star systems are not the place to look for planets with life.

Dear Merlin,

What are your thoughts on the theory that a star named Nemesis is circling our solar system and was responsible for killing off the dinosaurs? Is it plausible that this star is so small and faint that it has yet to be discovered and its gravitational pull has not been detected?

DIANA T. H. DUONG

BERKELEY, CALIFORNIA

It seems as though everybody is trying to find a way to kill the dinosaurs. The growing list of murder weapons includes an ice age, galactic gamma ray bursts, nearby supernova explosions, a giant meteor impact, and the Nemesis star.

Nemesis is a hypothetical member of the solar system that forms a binary star pair with the Sun. Its orbit is extremely elongated and it swings far enough away to perturb the orbits of countless comets in the outer solar system. These comets then rain down on the inner planets and wreak havoc upon Earth, making life wholly unpleasant for *T. rex* and friends.

The Nemesis star is a creative idea that just barely qualifies as a scientific theory: the star's orbit was artifically tuned to match the absence of evidence, and it makes no testable prediction. For example, no one has ever seen a companion star to the Sun, so the brightness and distance of Nemesis were constructed to make the star undetectable in the age of telescope searches, and the orbit period for Nemesis was set to match the long intervals between mass extinctions in the fossil record.

There may indeed be a Nemesis star, but at the moment Merlin is unconvinced, because the absence of evidence is not the same as the evidence of absence.

Dear Merlin,

How does our solar system stay in a fixed position within the revolution of the galaxy so that the stars of our constellations are always in their proper positions, century after century?

BEVERLY WALDEN
WIMBERLEY, TEXAS

You may have noticed when you drive down a road that the side of the road (anatomically called the shoulder) moves rapidly by the window, yet the trees and hills on the horizon move by much more slowly.

The Moon is even farther than the horizon. As you drive, the Moon moves by you so slowly that it doesn't appear to move at all; hence the childhood query "Why does the Moon follow me?" If you could drive at 55 miles per hour off Earth and into space for 100,000 hours, you would leave the Moon far behind.

Your question reworded might read, "Why do the constellations follow me?" The solar system moves quickly in orbit around the Milky Way galaxy, but the stars of the constellations are far enough away that they don't appear to move in a lifetime, or over many centuries.

In about 100,000 years, however, the familiar night sky patterns will have changed, so that the "new" constellations will be wholly unrecognizable to all the dead twenty-first-century stargazers.

Dear Merlin,

How do astronomers determine distances to the stars and planets?

DAVID COLLINS

TORONTO, ONTARIO, CANADA

*N*owadays we can bounce radio waves off the planets, time the delay, and compute the distance directly using the speed of light. Stars do not reflect radio waves very well (actually, they do not reflect any form of light very well), and they are much too distant for the return signal to be strong. A round-trip bounce to the nearest star would take over eight years.

The primary scheme by which we get distances to nearby stars is the parallax method, which is the basis for many secondary and tertiary methods that form the "distance ladder." By analogy to your anatomy, if you hold your thumb at arm's length and wink your left eye and then your right eye, you will notice that your thumb shifts right, then left against the background scenery. The shift will be twice as great if you extend your arm only halfway. By simple geometry, the amount of your thumb-shift and the separation of your eyeballs uniquely determine the distance to your thumb.

In astronomy, the background stars correspond to the background scenery, the diameter of Earth's orbit corresponds to the separation of your eyeballs, and the distance to the star in question corresponds to the distance to your thumb. Astronomers "wink" at the star by observing it at six-month intervals.

But what if you want the distance to the background stars? Nearly all other methods of distance determination, out to the farthest quasars, are based directly or indirectly on this method. The common terms "distance ladder" and "distance pyramid" metaphorically refer to this sobering fact.

Dear Merlin,

I've heard of red stars and blue stars and white stars. Are there any green stars?

CARLA THORNDIKE

KANSAS CITY, MISSOURI

No. But the reason isn't mysterious.

Most stars emit all visible wavelengths of light, which include red, orange, yellow, green, and blue. But if we look at the entire spectrum, visible and invisible, we find that the wavelength range from red to blue is quite narrow.

If a star's energy output peaks in the infrared, the star will look reddish, because it emits more visible red light than blue light.

If a star's energy output peaks in the ultraviolet, the star will look bluish, because it emits more visible blue light than red light.

If a star's energy output peaks within the narrow visible range, then approximately equal amounts of red through blue are emitted. Anytime this happens you have a white star—a veritable inverse rainbow.

Dear Merlin,

How do stars get their names? (Besides the North Star and the stars that are named after someone who discovered them.)

STEPHANIE PETZ

PITTSBURG, KANSAS

*F*or most constellations, the stars are lettered from bright to dim using the Greek alphabet (alpha, beta, gamma, delta, etc.). They are further identified by the constellation's genitive form. For example, the brightest star in the constellation Centaurus is Alpha Centauri, and the brightest star in Canis Major is Alpha Canis Majoris.

Some stars also have common names. Alpha Canis Majoris, as any amateur astronomer knows, is Sirius, the brightest star in the nighttime sky. Other famous stars include Alpha Lyrae, better known as Vega, and Alpha Ursa Minoris, the North Star.

Unlike comets and asteroids, only a few stars are named for their discoverer. The best-known example is Barnard's star (named for the nineteenth-century American astronomer Edward E. Barnard), in the constellation Ophiuchus. At a speed of 10 arc seconds per year (a degree every 360 years), Barnard's star is among the fastest-moving stars in the sky.

Other stars have names derived from mythology. For example, the name of the brilliant red star Antares, the brightest star in the constellation Scorpius, literally translates to "against Mars," because the star was perceived as a competitor to Mars's redness in the sky.

Apart from these naming schemes, tens of thousands of visible stars in the nighttime sky—the bright ones and the dim ones— are boringly numbered in sequence of their right ascension and listed in the *Smithsonian Astrophysical Observatory Star Catalog.*

Dear Merlin,

Are Mizar and Alcor in Ursa Major really binaries? What is their period? In the almost fifty years I have watched them, they never seem to have moved.

ALEXANDRA ACKERMAN
BLOOMINGTON, INDIANA

*I*f you had seen Mizar and Alcor move, you would be no less than 100,000 years old.

As you know, these middle stars of the handle of the Big Dipper are detected easily with the unaided eye. Mizar (the brighter of the two) and Alcor are not near enough to each other to consider them an official orbiting binary pair. They are, however, part of the loosely bound Ursa Major star cluster, which contains, among others, five of the seven Big Dipper stars.

If you like double stars, you needn't be upset. Alcor itself is a double star. Mizar itself is a double star. And both stars that are the Mizar double are also double stars.

Dear Merlin,

A recent article I read discusses the triple Centauri star system. Are triple star systems common in our galaxy? It would be great to live somewhere and watch three sunsets and three sunrises every day.

EDMUND ROACHE, M.D.

WATERTOWN, NEW YORK

$\mathcal{M}$ost multiple star systems in the galaxy are binaries, but triple star systems are not rare.

But before you make travel plans, consider that planets around multiple star systems tend to have wildly chaotic orbits. Their gravitational allegiance shifts continuously from one star to another and inevitably leads to their getting ejected from the system. In Merlin's opinion, romantic triple sunsets do not compensate for life on a jack-in-the-box planet.

Dear Merlin,

Do stars and other space objects emit sounds? If we were close enough, would we hear the Sun's explosions?

NEIL LANE

PHOENIX, ARIZONA

Yes. Yes.

The Sun's atmosphere is quite a noisy place, with its gurgling turbulence and its explosive flares. But there is one complication. If you were close enough to hear the Sun, you would be close enough to be vaporized, so the audibility of the Sun would no longer be a concern of yours.

Outside of the Sun's atmosphere, throughout interplanetary space, there is no gas or material medium to carry sound waves. This emptiness is what renders a deep cosmic silence upon the universe. Blockbuster science-fiction films tend to ignore this fact.

Dear Merlin,

Why are planetary nebulas called planetary nebulas? Don't they have nothing to do with planets? Why don't you suggest a better name?

JOHN BRAND

INDIANAPOLIS, INDIANA

$\mathcal{M}$isnomers are quite common in astronomy. You have hit upon one of them. What follows is the story of planetary nebulas.

Once upon a time, in the early days of the telescope, astronomers were eager to classify all objects that were observed. An obvious distinction between planets and stars (as seen through a telescope from Earth) was that planets looked like little disks and stars looked like little points of light. A next step was to classify the fuzzy, nebulous things. Some fuzzy things were spiral-shaped, yet they remained fuzzy in a telescope. They were called spiral nebulas (later discovered to be spiral galaxies). Others were elliptical. They were called elliptical nebulas (later discovered to be elliptical galaxies). Some nebulas had no special shape other than what they happened to resemble in the mind of the discoverer. The Horsehead Nebula (interstellar dust) and the Crab Nebula (a supernova remnant) are part of a very long list. But one category of nebulosity looked like little disks. The only other little disks astronomers knew about were planets. Thus was born the name "planetary nebulas." Planetary nebulas are now known to be the spherical envelopes of dead red giant stars that have gently lifted their outer layers and laid bare their small hot core—the white dwarf. And the nebulas expanded happily ever after.

Perhaps you will now forgive planetary nebulas, and all those who use the term. But if you still seek a new name, Merlin votes for "puffball nebulas."

Dear Merlin,

How common are supernovas?

GREGORY MILLER

SAN DIEGO, CALIFORNIA

Supernovas, the explosive, in-your-face, violent, neighborhood-disrupting death call of high-mass stars, are actually quite common throughout the universe. From Earth, many dozen are discovered each year among catalogued and occasionally uncatalogued galaxies. They are lettered annually in sequence of discovery and thus become endowed with unexciting names like SN1983K or SN1991M. When more than an alphabet's worth of supernovas are discovered in a year, the letters are doubled up.

The most celebrated supernova of modern times, SN1987A, exploded in the Large Magellanic Cloud, which is one of the Milky Way galaxy's closest neighbors. The event was visible to the unaided eye for months—a treat that occurs only three to four times per millennium. Unlike all other supernovas ever observed, SN1987A landed the cover story of *Time* magazine.

Dear Merlin,

I heard or read somewhere about a famous Chinese supernova. Would you please clarify my memory of this event?

GREGORY MILLER

SAN DIEGO, CALIFORNIA

The brightest recorded supernova in your Milky Way galaxy to be visible from Earth was seen and recorded by the Chinese during the Sung Dynasty on the fourth of July in A.D. 1054. Europe was in the Dark Ages and made no record of the event. The supernova is still celebrated annually with fireworks by astronomers in America.

The remnant of the explosion remains visible as a crab-shaped nebula with a dense, rapidly rotating neutron star at its center. In keeping with the official naming scheme for cosmic nebulosity, this supernova remnant is called the Crab Nebula.

Dear Merlin,

Does radiation from a supernova many light-years away have any effect on life here on Earth? If it does, what symptoms would it give the human population? That is, what kind of radiation sickness?

D. FRY

PHILADELPHIA, PENNSYLVANIA

*M*erlin does not know how many light-years you mean when you say "many light-years." If "many" means several hundred thousand (like the well-publicized SN1987A), then life on Earth is safe. Radiation intensity drops significantly over these distances.

If "many" means up to one hundred light-years, then the high-energy ultraviolet rays, X-rays, and gamma rays emitted by the supernova would bathe Earth with such intensity that the molecules of human life would decompose. Everyone would contract the advanced stage of radiation sickness that is commonly known as death.

No need to worry. There is no supernova candidate within this distance.

Dear Merlin,

If the first generation of stars contained only hydrogen and helium, and a subsequent generation of stars built upon what was produced to create our current elements, and if we are now in a third generation of stars, then what can we expect from fourth-generation stars? Have any been detected yet?

THOMAS L. EFFINGER

POSEYVILLE, INDIANA

In the early concept of stellar population types, there was a first generation of stars and then a second generation and so forth. This idea has recently evolved to recognize a continuous range of population types in regions such as the disk of spiral galaxies, where the formation of stars and the enrichment of heavy elements from supernovas proceed continuously over time.

The element iron makes an excellent tracer of all elements heavier than hydrogen and helium. The stars with the lowest known heavy element abundance, found in the halo of the Milky Way, have about one ten-thousandth the amount of iron found in the Sun. The stars with the highest known heavy element abundance, found in the bulge of the Milky Way (near the galactic center), have over three times as much iron as is found in the Sun.

Dear Merlin,

Approximately what percentage of the stars we see in our sky are members of our own galaxy?

MIKE GLEN
AUSTIN, TEXAS

If you have normal human vision, Merlin can give you an exact total: 100 percent of the stars you see with the unaided eye belong to your Milky Way galaxy.

Only three objects that do not belong to the Milky Way are visible. They are Merlin's home (Andromeda galaxy, 2.2 million light-years away) and two nearby dwarf galaxies (the Large and Small Magellanic Clouds, each about 160,000 light-years away). With a telescope, you can detect individual stars in these galaxies and in many other nearby systems. With the largest of telescopes, the billions of galaxies that compose the universe come into view.

You didn't ask, but 100 percent of everything detected by the world's largest telescopes belong to our one and only universe.

Dear Merlin,

It seems to me that estimates for the number of stars in the universe are a little tenuous. If the uncertainties in the age of the expanding universe (and hence its diameter) lie somewhere between 8 and 15 billion years, it would seem that your estimate of the size of the universe (and hence the number of stars) must range by a factor of eight or so.

I don't suppose that a factor of eight much matters—another zero more or less being hardly noticeable with such large numbers—but it does suggest the need to examine other assumptions that go into your calculation. On what basis, for instance, do you infer that the universe is a sphere? Why not a cube or, better yet, a shoe? I think I prefer the cube, if only because a cube would pack so much better if it turns out that there are other universes to calculate.

Randolph Briwn
Brimley, Michigan

$\mathcal{M}$erlin's estimate for the number of stars in the universe is about one sextillion, or 1×10^{21}, which is a middle-of-the-road value. If the volume were ten times larger, the number of stars would become 1×10^{22}. Suppose the universe were one tenth the volume that we previously thought. Then the number of stars becomes 1×10^{20}. These factor-of-ten uncertainties sound large but are in fact small when compared with the size of the original number. We should all rejoice that the uncertainty in either direction is not a factor of a thousand, or a million.

The observable universe forms a spherical volume around us and is bound in all directions by either the beginning of the universe or the maximum distance light could have traveled during the history of the universe. The calculation is simple: If the universe is 12 billion years old, then the radius of the sphere is 12 billion light-years.

There is otherwise no evidence to suggest that the universe is in the shape of anybody's shoe or a cube—in spite of how convenient it would be to stack multiple universes on your pantry shelf.

IV

GALAXIES,
NEAR and FAR

Dear Merlin,

I often look at the Milky Way stretched across the night sky. Where is the center of our galaxy? Is it star-packed like the middle of the Andromeda galaxy? If not, can an amateur astronomer find it in some constellation?

GREGORY MILLER

SAN DIEGO, CALIFORNIA

$\mathcal{M}$ost spiral galaxies (the Milky Way and Andromeda included) are star-packed in their center. Often, however, dense clouds of dust and gas obscure the view.

If you are a hardcore amateur astronomer, then you can find the galactic center at right ascension 17 hr 45.5 min, declination: −28° 56′ (epoch: 2000). If you are a humble stargazer, then you will find the galactic center if you gaze 3 degrees (about one finger-width at arm's length) to the west of the teakettle's spout in the constellation Sagittarius.

If you only look up occasionally on the weekends, then the galactic center will be in front of your face if you look due south a few hours after sunset in late September and early October.

Dear Merlin,

I have read that our solar system lies half of the distance from the center to the edge of our Milky Way galaxy. As I gaze at Sagittarius toward the galactic center in the summer, I wonder, how is the plane of our solar system oriented to the plane of the galaxy? Are we north or south of the galactic plane?

MICHAEL FLOCKE

AUSTIN, TEXAS

The entire solar system is tipped nearly 90 degrees to the galactic plane and can be found nested in it along with more than 100 billion other stars.

Dear Merlin,

I have this large posterlike photograph of the great galaxy in Andromeda, which is said to be the closest large galaxy to our own Milky Way and similar in shape. What I would like to know is, what are those stars spattered all over the picture? Are there thousands of stars between Andromeda and our galaxy, or are they part of our own Milky Way?

LESTER DUBACH HELLMAN
PRESCOTT VALLEY, ARIZONA

Les wants not to believe a fallacy.
"My photo is spattered with stars, you see!"
Merlin says to him,
"The stars are within
The bounds of the Milky Way galaxy."

Dear Merlin,

Are there any stars that are not part of a galaxy?

MARK MASON

AUSTIN, TEXAS

A single intergalactic star would be very hard to notice unless, for example, it exploded as a supernova (and thus became a million times brighter). Among the thousand or so supernovas that have ever been discovered, about a dozen have exploded suspiciously distant from their host galaxy. These form the best evidence available for the existence of intergalactic stars.

A question remains, however: How did they get there?

Dear Merlin,

    How many galaxies are visible to the naked eye?
    AMANDA SMITH
    LAUREL, MARYLAND

Of the billions and billions and billions of galaxies in the universe, only four are visible to the unaided eye:

1. The Milky Way galaxy (you are in it right now).
2. The great spiral galaxy in the constellation Andromeda (also known as the Andromeda galaxy—Merlin's home).
3. The Large Magellanic Cloud (a small satellite galaxy to the Milky Way).
4. The Small Magellanic Cloud (an even smaller satellite galaxy to the Milky Way).

Dear Merlin,

Do all spiral galaxies rotate in the same direction?

FLETCHER C. PADDISON

AUSTIN, TEXAS

*E*vidence gathered from spiral galaxies across the universe shows that their disks are oriented at all angles to each other, which means you are as likely to find two that rotate in the same direction as in opposite directions.

If you like rotational mayhem, you may be intrigued by recent data from nearby spirals that reveal counterrotating disks of stars *within the same galaxy.*

Dear Merlin,

Since the time required for a star to complete one orbit around the galaxy is proportional to the distance from the center (i.e., the farther away it is, the longer it takes to orbit), why do the arms in a spiral galaxy not get wound up? Also, why do galaxies have arms, and is it always two? Also, why do stars in galaxies have (apparently) circular orbits while virtually everything else has elliptical orbits?

DAN NEWLAND

ALAMEDA, CALIFORNIA

Spiral arms are relatively permanent features that rotate at different "pattern speed" when compared with the rest of the galaxy. They are regions of slightly higher gravity that occur naturally in flattened systems, such as spiral galaxies. Best described as a density wave, an arm is where stars and clouds slow down in their orbit as they pass through this gravitational disturbance.

The phenomenon is similar to the effect of rubbernecking on the flow of traffic—cars eventually pass the trouble spot, but they are slowed in the process.

Many galaxies, including your Milky Way, have discrete arms, while others are flocculent. Even though stellar orbits are often approximated by circles, most have complicated shapes that look like neither circles nor flattened ellipses.

Dear Merlin,

I realize that Earth lies in a rather flat region of our galaxy. I wonder what the night sky would look like if we were on a planet in an elliptical galaxy. Let's say the planet was two thirds of the way from the galactic center, just as Earth is two thirds of the way from the center of the Milky Way. What kind of sky would the naked-eye observer see in this situation?

KAREN D. GUETZOW

SAN BENITO, TEXAS

*L*et's consider the titanic elliptical galaxy M87 in the Virgo cluster of galaxies. If Earth were embedded among the trillions of stars that compose this galaxy,

1. The entire nighttime sky would be aglow with starlight.
2. One might find it difficult casually to distinguish twilight from midnight.
3. The sky brightness would be so high that you would have no optical information on the contents of the rest of the universe.
4. Radio telescopes would provide one of the few means by which external galaxies could be detected.
5. There would be so many stars that you could find constellation patterns that looked like almost anything.

And you probably would have named the galaxy Milky Soup rather than Milky Way.

Dear Merlin,

You say that elliptical galaxies do not rotate. If that is true, why don't they collapse? Could it be that they are collapsing and we do not detect it?

JAMES BERRY

TUSTIN, CALIFORNIA

*M*erlin never said that elliptical galaxies do not rotate. But they rotate very slowly when compared with the disks of spiral galaxies. Like bees in a swarm, their puffed, elliptical shape comes from the collective boundary of all stars in their orbits.

Dear Merlin,

What are the largest galaxies? Are they the elliptical or the spiral type?

ED CRASTON

SEATTLE, WASHINGTON

The gas-rich, star-forming spiral galaxies may be pretty, but elliptical galaxies hold all the records. The largest, the brightest, the smallest, and the dimmest known galaxies are all some variety of elliptical galaxy.

The largest galaxies in the elliptical family tend to loom most often near the centers of crowded galaxy clusters. Many have multiple nuclei, which leads to the suspicion that whole galaxies have been cannibalized.

These galactic cannibals can easily accumulate over ten times the mass of the entire Milky Way.

Dear Merlin,

Please yield some information on the object "galaxy cluster" marked rather prominently in sky charts near the stars Alpha and Beta Corona Borealis. I've tried repeatedly to capture it in both binoculars and telescopes, but to no avail. Is it a clustery system of stars belonging to our own galaxy?

H. C. SUEHS

FESTUS, MISSOURI

*M*erlin likes tenacious observers. But sometimes you just need a bigger telescope. The Corona Borealis cluster is a rich assortment of galaxies, not stars. Over four hundred of them are packed in an area of the sky smaller than the full moon. Unfortunately for you, it's over a billion light-years away, so the brightest galaxies top out at about 16th magnitude, which is ten thousand times dimmer than the limit of the unaided human eye.

Dear Merlin,

I read that galaxies are the largest definitive structures in the universe. How certain are we of this? Couldn't it be true that stars are to galaxies as galaxies are to some other definitive structure, such as superclusters?

IAN M. CAREY

FOLSOM, CALIFORNIA

Galaxy superclusters (or for that matter superduperclusters) do not tend to be "definitive" structures. For most of them, the typical space velocities are not high enough for any galaxy to have crossed the width of the cluster in the age of the universe.

Regardless, less than 10 percent of all known galaxies are found in clusters of any size. It is therefore honest and fair to say that stars compose galaxies and that galaxies compose the universe.

Dear Merlin,

I read about how researchers carefully measured the distances to the nearest galaxies and produced a three-dimensional map of one small slice of the sky. The galaxies seemed to be ordered into ringlike shapes (bubbles?) surrounded by vast voids. These distances, as I understand it, also represent a shift backward in time. Has anyone adjusted this 3-D map to show how the shape changes from the date of the oldest (farthest away) galaxy to the present?

ALEX VRENIOS

AUSTIN, TEXAS

The farthest galaxy in one of those 3-D maps is typically hundreds of millions of light-years away, moving with a space velocity of hundreds of miles per second. The light we see left such a galaxy hundreds of millions of years ago.

Across this time, however, the galaxy has moved a distance no greater than about ten of its own diameters. Consequently, to "reverse the clock" is to produce a slightly different picture—but not one that is particularly more interesting than what you started with.

Dear Merlin,

I understand that in the big bang, all galaxies are receding from one another, with the farther galaxies receding faster. How is it that some galaxies are moving toward the Milky Way, such as the Andromeda galaxy?

WALT PERKO
SAN FRANCISCO, CALIFORNIA

*A*ll galaxies have some motion that is separate from the overall expansion of the universe. For the distant galaxies (farther than about 100 million light-years), a galaxy's velocity from the expansion of the universe will dwarf the galaxy's independent velocity.

Only for nearby galaxies, where the expansion rate of the universe is relatively small, will you find the velocity of an individual galaxy that is comparable to the velocity from the expanding universe. Indeed, Merlin's home, the Andromeda galaxy (a mere 2.2 million light-years away), shows a velocity toward the Milky Way of about 100 kilometers per second.

Dear Merlin,

I read recently that the presence of quasars indicates extremely distant galaxies receding from us at tremendous speeds. Would it not be reasonable to suppose that these distant galaxies might be, relative to us, past the point where the big bang occurred?

AL CRAIG

ESTILL, SOUTH CAROLINA

The idea that one can see "through" the big bang to negative time is intriguing. Seeing through something (a window, air, space, the big bang) requires photons of light to move from the source to the observer unobstructed.

But there was an epoch in the hot early universe when photon paths were severely obstructed. Transparency was poor in much the same way that frosted, translucent glass hides what's behind it. All visible knowledge of the structure and contents of the universe begins at this epoch, not at the big bang.

This photon "barrier" occurred at a redshift of $z \approx 1500$, which corresponds to a time when the universe was several hundred thousand years old. The most distant known quasars are found between redshifts $z = 4$ and $z = 6$, when the universe was about a healthy billion years old.

V

A Look
at the
Day Sky

Dear Merlin,

When we look at the Sun, do we see it where it is or where it was? If we see it where it was, then I presume the objects we see in the heavens must not be where we see them. NASA really should be on top of this point, because if they shoot a probe to one of those heavenly bodies, it may not be where they claim.

JACK R. BIRCHUM
LANCASTER, TEXAS

Because light takes time to get from one place to another, you (and everybody else) see things not as they are but as they used to be, and not where they are but where they once were. The light from this page is a billionth of a second old by the time it reaches your retina—the page probably hasn't moved very much in that time. But the light from the Sun is five hundred seconds old by the time it reaches Earth, and the light from Jupiter is about forty-five minutes old by the time it reaches Earth. These objects are not where we see them.

If all NASA did was aim and shoot, there would probably be many stray probes. Orbit corrections en route are common, and they are each sent in advance to allow for the time it takes the radio signals (traveling at the speed of light) to reach the probe.

Dear Merlin,

If the ozone layer of the atmosphere has a hole in it, will the amount of radiation from the Sun increase, decrease, or will it stay the same?

VICKIE PADRON

STAMFORD, TEXAS

$\mathcal{M}$erlin assumes you refer to the amount of the Sun's radiation that reaches Earth's surface.

Ozone in the atmosphere absorbs ultraviolet radiation. This variety of light can severely damage the eye's retina cells (commonly called blindness), darken the skin (commonly called getting a tan), induce malignant melanoma (commonly called skin cancer), and promote premature toughening and wrinkling of the skin. People with light skin, blond hair, and/or blue eyes are at the greatest risk.

A hole in the ozone layer would locally render Earth's atmosphere transparent to ultraviolet radiation and therefore would locally increase the incidence of ultraviolet-related health problems.

Dear Merlin,

How fast is the Sun spinning?

J. CANTLEY

SALINAS, CALIFORNIA

The gaseous Sun does not manage to "keep it all together" as it rotates.

Its equator spins once in about twenty-five Earth days, while near the poles it spins once in about thirty-one Earth days. This produces a shearing effect on the Sun's atmosphere that distorts the solar magnetic field and drives the development of sunspots, solar flares, and other surface activity.

Dear Merlin,

My father and I follow the spots on the Sun almost every day and we have a question. A book in the library says that sunspots only happen on the very top or bottom of the Sun and that they revolve around and around it. But we've seen lots of them near the center and they seem to move in circular patterns on the face of the Sun. Why is that?

AMELIA MYRDEL SHULTS-HILL
COLORADO SPRINGS, COLORADO

$\mathcal{M}$erlin does not know what library book you used, but it should not have said that sunspots appear at the very top or bottom of the Sun. Sunspots are found, without exception, between +50 and −50 degrees solar latitude. A typical sunspot will take about thirteen days to cross the Sun's face. In addition, sunspots are locked into the rotation of the Sun, so they should not trace circles or exhibit other playful behavior.

Merlin suspects that you and your father observed the Sun by projecting its image onto a wall or screen—the safest of all methods. Because of the Sun's arc across the sky, its projected image will systematically turn, so that over six hours the Sun's axis will have tipped by 90 degrees. If you didn't know this, the sunspots would most assuredly appear to trace little circles on the Sun's disk.

Dear Merlin,

Suppose I get a clear, unobstructed view of a flat western horizon from my front porch. On the winter solstice I note the point on the horizon where the Sun sets. Six months later, on the summer solstice, I do the same thing again. What would be the angular separation of these two points? What, if any, effect would my latitude have on this measurement?

JAMES LUSK

BERKELEY, CALIFORNIA

*R*egardless of your latitude in the Northern Hemisphere, the Sun rises (and sets) its farthest south of east on the winter solstice. Each day thereafter, the Sun rises (and sets) farther and farther north along the horizon until the summer solstice, when it rises its farthest north of east. The maximum separation ranges from 47 degrees, if you are an equator-dweller, all the way up to a full 180 degrees, if you live on the Arctic or Antarctic Circle (66½ degrees north and south latitudes).

For the record, the old adage "The Sun rises in the east and sets in the west" is only true on two days of the year: the spring and the autumn equinox, where it is true for every spot on Earth except the poles, where, of course, there is no east or west.

Dear Merlin,
  What is the "green flash"?
HANK RENFIELD
CHICAGO, ILLINOIS

This phenomenon, one of several in astronomy whose name sounds like a comic-book hero, is a rare atmospheric effect seen during the last moments of the setting (or rising) Sun.

If you leave urban Chicago to get a clear view of the western horizon, and if the atmosphere is layered and stable (no big puffy clouds, storms, or tornadoes), and if you have nothing better to do than watch every possible sunset, then one day you may see the last upper edge of the setting Sun turn green.

The thickest line of sight through Earth's atmosphere is toward the horizon. In stable air, the atmosphere can act like a prism, vertically splitting the sunset light into the familiar rainbow colors: red, yellow, green, blue, indigo, and violet. As the setting spectrum slides past your view, the blue-indigo-violet light is mostly scattered around the sky, and the red-orange-yellow light goes unnoticed against the setting Sun. What you notice is a flash of green, which has been aptly named the "green flash."

Dear Merlin,

I understand that if Earth revolved around two suns, objects and beings would cast a double shadow. But wouldn't a planet's orbit around two stars together be impossible because of the greatly fluctuating level of gravity in such a busy system?

FLO BEAUMON

SEATTLE, WASHINGTON

There is nothing bad or impossible about a fluctuating gravity. Earth, for example, feels a constantly changing force of gravity as it traces an elliptical orbit around the Sun.

The path of any planet around one, two, five, or a million stars will always be in response to the sum of all gravitational forces. What makes these many-body systems interesting is that orbital paths tend to become chaotic, and often planets get ejected.

If you are lucky enough not to get ejected, then the more relevant question is, can life survive on a planet with a wild and eccentric orbit? The enormous change in distance to the various stars may produce far too great a temperature variation to sustain stable life forms. On such a planet, people probably would care more about their survival than about the appearance of their shadows.

Dear Merlin,

Why can we sometimes view our Moon in the daytime?

DAVID SORRELLS

EAST POINT, GEORGIA

The Moon rises in the daytime just as often as it rises at night. In fact, the Moon can be spotted in the daytime sky on nearly twenty-four days out of the twenty-nine and a half days of the lunar cycle. For example, the last quarter moon (which doesn't rise until about midnight) is high in the sky when you wake up at sunrise. By the time you brush your teeth, the Sun is fully risen. And by the time you walk out the front door, the moonlight has trouble competing with the sunlight for your attention. Often the Moon just goes unnoticed. For this reason, many people don't even know that the Moon also "comes out" in the daytime.

Dear Merlin,

This filler appeared in the Temple Daily Telegram on August 13, 1993: "DARKNESS DEVELOPS: In 1780 a mysterious darkness enveloped much of New England and part of Canada in the early afternoon; the cause has never been determined." Do you have any further information about this event, or know where I might learn more?

CONSTANCE LEE

TEMPLE, TEXAS

Not all mysteries are mysterious. There was a total solar eclipse across upper New England above a cloud-covered sky at midday on October 27, 1780.

Dear Merlin,

Why are some total solar eclipses longer than others? I've heard that some are as short as a minute and others are as long as seven minutes.

VANESSA GREEN

GALVESTON, TEXAS

The popular song lyric "I'm being followed by a moon shadow" would be true only if you traveled at moon-shadow speed (over 1,300 miles per hour) ahead of the exact eclipse path as it passed over Earth's rotating surface. If you slowed down a bit and let the shadow catch up, you would get to view the unilluminated near side of the Moon as it covered the Sun. This song-inspired scenario would sustain a total solar eclipse for hours.

To witness a total solar eclipse the normal way is simply to stand where you know the Moon's shadow will pass. If the Moon's shadow on Earth is big, then it will take longer to pass over your location. To get a big shadow, you need a big Moon and a small Sun.

The Sun looks smaller in the first week of July than at any other time of year, since this is when Earth, in its elliptical orbit, is farthest from the Sun (aphelion). The Moon will look its biggest, of course, when closest to Earth (perigee) in its monthly elliptical journey. Combine aphelion and perigee and you have a whopping total eclipse in excess of seven minutes, which is why, for example, the total solar eclipses of 30 June 1973 and 11 July 1991 were so long.

Other combinations of Earth-Sun and Earth-Moon distances will give shorter total eclipses, with an average of between two and four minutes.

Dear Merlin,

After reading warnings about solar eclipses, I am concerned about sunlight. My doctor tells me to stay out of the sun from 10 A.M. to 3 P.M. So I swim outdoors at about sundown, using the backstroke. If I should look at the setting Sun for a minute or so, is this dangerous to my eyes?

RUTH B. PECK
FALLS CHURCH, VIRGINIA

The official word is that you should never look directly at the Sun without specially designed protective filters.

The unofficial word is that occasionally semitransparent clouds pass in front of the Sun, or a sunset occurs through haze or smog. When this happens, the Sun's intensity is significantly reduced. If you happen to glance at it, you needn't run to your eye doctor. Note that a glance translates to seconds, not minutes.

Dear Merlin,

I have seen one total solar eclipse from the replica of Stonehenge above the Columbia River in Washington some ten or eleven years ago. It was so spectacular I hope this was not a once-in-a-lifetime event. Will I be able to witness another total solar eclipse from near my new home at 7° E, 46° 15' N anytime soon?

JAMES F. POPE

LEYSIN, SWITZERLAND

*F*rom your charming Alpine village of Leysin you are almost at yodeling distance from the total solar eclipse on the morning of 11 August 1999. That eclipse path touches down in the North Atlantic and heads east, passing just south of London. It then cuts a swath across France and continues onward through Freiberg and Munich, Germany. Any spot about 200 kilometers north of Leysin should do just fine.

Merlin strongly advises that you put this eclipse date on your calendar, because in all likelihood you will be dead for the next total solar eclipse, in the late afternoon of 14 June 2151.

Dear Merlin,

I observe the Sun often through solar filters with my 60mm refracting telescope. On March 20, 1992, I saw an unusual sight: a very small round object passed in front of the Sun. While the object was in transit I began to count "one M-i-s-s-i-s-s-i-p-p-i, two M-i-s-s-i-s-s-i-p-p-i," etc., to gauge its transit time, but because of my excitement I stopped before it had finished crossing. I counted to twelve but estimate the total transit time to be as much as thirty seconds. The object did not cross at the widest part of the Sun's disk. Although I looked through a variety of books, I have been unable to find any information on what the object might be. I would appreciate any comments or insight that you might be able to give.

    EDWARD J. PRAITIS

    REDMOND, WASHINGTON

*R*esidents of Mississippi are no doubt pleased that their state's eleven letters assisted your observation.

The space between Earth's surface and the rest of the universe is filled with all sorts of objects that can pass in your field of view—birds, airplanes, flying superheroes, satellites, meteors, comets, asteroids, and planets. Neither Mercury nor Venus transited the Sun in 1992. While asteroids and meteors cannot be excluded, Merlin suspects that you witnessed the transit of a geosynchronous communications satellite. These satellites revolve around Earth over the equator at the same rate that Earth rotates. Geosynchronous satellites thus appear stationary over a chosen spot on Earth, yet will appear to move across the sky relative to the Sun.

You can only view the transit of a geosynchronous satellite when the Sun crosses the projection of Earth's equator on the sky. In other words, it could only happen during the spring or fall equinox. Your observation was made on the spring equinox of 1992.

Dear Merlin,

Recently, when I was visiting Sanibel Island, Florida, the Sun rose at 7:17 A.M. yet I kept observing Venus and Jupiter until 7:40 A.M. Is it unusual to see planets with the naked eye by daylight?

ALEXANDRA ACKERMAN

BLOOMINGTON, INDIANA

*I*t's relatively easy (and not unusual) to follow Venus and Jupiter long into the dawn sky with the unaided eye. You do not even need to be on vacation in Florida.

If you want a real challenge, try to *find* Venus, Mars, Jupiter, and Saturn in broad daylight. At their brightest they can all be seen with an amateur telescope—if you know where to look.

VI

A Look

at the

Night Sky

Dear Merlin,

Why does the Moon come out only at night?

ROSEMARY TAGGART

LANSING, MICHIGAN

*P*eople have been confused ever since biblical Genesis said, "And God made two great lights: the sun to rule the day, and the moon to rule the night."

Let it be known that half of the time you have ever looked at the night sky, the Moon was *not* there. The saying "The Moon comes out at night" ought to be reworded, "At night, if the Moon is out, you may notice it."

If you look carefully, you will notice the Moon in the daytime sky just as often as in the nighttime sky. The exact location in the sky and time to see the Moon will depend on the Moon's phase.

Dear Merlin,

Why does the Moon look bigger on the horizon? Is it closer? Is it the atmosphere? Or is it psychological? I have heard all three reasons.

RICK LOWE

AURORA, COLORADO

The Moon is not any bigger on the horizon than elsewhere. Careful measurements of its width there and higher in the sky will yield the same results. Since the Moon does in fact look bigger on the horizon, we must defer to a psychological explanation.

Remarkably, there is still no agreement among Moon-on-horizon experts as to why the Moon looks bigger on the horizon. It seems likely, however, that trees and buildings (silhouetted against the rising or setting Moon) confound the depth cues that the human mind uses to perceive the Moon's size. On the open ocean, where trees and buildings are rare, the effect is reduced significantly.

Another way to remove the faulty depth cues is to bend over and view the Moon through your legs. (It's best to do this when nobody else is around.) The trees and buildings are upside down and less useful to the mind's perception of depth. Once again, the effect is reduced significantly.

Dear Merlin,

The description of the phases of the Moon has me confused. If one accepts the "full moon" term, then wouldn't it be more accurate to label the first and third quarters as the "first and second halves"? Alternatively, if one accepts the "first and third quarter" terms, wouldn't it be more accurate to label the full moon the "half moon"?

RON GIFFIN

CINCINNATI, OHIO

Yes. Yes. But sometimes tradition and fashion weigh more than rational thinking. As such, we must accept some things in life without getting upset about them.

Traditionally, the "full" in "full moon" refers to the Moon's appearance, whereas the "quarter" in "first and third quarter" refers to the Moon's position in its four-part monthly cycle. If you wish to refer to the quarter phases as first and second half-moons, Merlin guarantees that everybody will still know what you mean.

Dear Merlin,

You once stated that full earth seen from the Moon is over fifty times brighter than full moon seen from Earth. What is your reference source, or how did you obtain this information?

LOWELL L. KOONTZ

ALEXANDRIA, VIRGINIA

*M*erlin's references are in many cases the discovering scientists themselves—from Trog & Lodyte (the cave couple who invented the wheel) to the Nobel laureates of today. In other cases, such as the subject of your question, Merlin's abacus proves to be quite useful.

If you place Earth and Moon side by side and observe them from space (you may need a spacesuit for this experiment), you will readily see that Earth blocks nearly fourteen times as much sky as the Moon. In other words, Earth has nearly fourteen times the reflecting area. Additionally, Earth (clouds included) reflects, on average, four times better than the Moon. Some fast abacus strokes reveal that full earth is "over fifty times brighter than full moon seen from Earth" ($14 \times 4 = 56$).

If you like official words, then the term for reflectivity is "albedo." Some comparison albedos for visible light: matte black paint is close to 0.0, the Moon is 0.07, Earth averages 0.30, and appliance white paint is close to 1.0.

Dear Merlin,

I have a question about blue moons. Is it true that occasionally the Moon turns blue? If so, what makes it happen?

PHYLLIS LINDLEY

STONYFORD, CALIFORNIA

*O*nce in the twentieth century, the Moon actually appeared blue in the sky. In 1950, a 250,000-acre forest fire in Canada released particles to the upper atmosphere that circled Earth eastward. The Moon was blue on the night of September 26 as seen from Great Britain.

Other events can also alter the Moon's color. During a total lunar eclipse, the full moon can become anything from totally dark to deep yellow to deep burgundy. After the 1883 volcanic eruption of Krakatoa (which, incidentally, is *west* of Java), the Moon appeared green.

Dear Merlin,

The year 1990 ended with a full moon on December 31. When will be the next time the blue moon falls on the last day of the year?

KENNETH W. GEDDES

COLORADO SPRINGS, COLORADO

The Gregorian calendar is not based on the phases of the Moon. It is therefore one of the few calendars in the world that can have a full moon on New Year's Eve. That night's moon is possibly the most noticed of the year—the full moon *and* party people are up all night.

The next such occurrence is 31 December 2009, and the next one after that is 31 December 2028.

Dear Merlin,

Why is it that the Moon and stars move through the sky at the same rate in any given night? That is, why doesn't the Moon move through the night sky at a faster (or slower) rate than the stars, since it is so much closer than the stars? Its position relative to the stars on a given night does not seem to change.

I suppose that the stars' apparent movement is due solely to Earth's daily rotation. But the Moon, a close celestial body, is orbiting Earth in addition to Earth's rotation. So it seems quite a coincidence that it ends up moving with the stars.

PAT SUTTON
DENVER, COLORADO

Look harder.

The Moon moves "backward" (west to east) against the background stars by an amount equal to its own diameter every fifty-five minutes.

Dear Merlin,

As a stargazer from Down Under, I often must grapple with the orientation of the star maps (they show far too much northern sky for my liking), and in doing so I constantly question the relative locations of the ecliptic and equator shown thereon. My question is, shouldn't the ecliptic be shown below (south of) the equator during the time of southern summer (northern winter)? The ecliptic, as I understand it, is the mean plane of the planets' orbits around the Sun, which should be constant. Hence, as the Southern Hemisphere becomes exposed to the Sun during the southern winter, the Earth's orbital tilt should have the equator above (north of) the ecliptic. If not, where am I going wrong?

MALCOLM R. MCKELLAR

MOSSMAN, AUSTRALIA

G'day, mate. No need to fret. The ecliptic is defined as the path against the background stars that the Sun takes throughout the year during Earth's orbit. As you correctly noted, the other planets do not stray too far from this path.

The ecliptic, as it is drawn in monthly star charts, will always appear as a segment of where the Sun isn't; otherwise it would be daytime and stargazing would be uninteresting.

In other words, Earth's seasons are determined by the location of the Sun on the ecliptic, not where you see the ecliptic drawn on a nighttime star chart.

Dear Merlin,

All the planets have times when they disappear from view for many nights. How did the ancients know that they were looking at the same planet when it reappeared?

MERWIN LUCAS

GLENDALE, CALIFORNIA

The ancients of three millennia ago did not have modern nightly distractions like primetime television and video rentals. Nor were they likely to be in such a hurry that they never looked up.

If you follow the weekly path of an "evening" planet in the sky as it nears the Sun's glare, you can easily infer that it will eventually reappear off to the other side of the Sun. To test your hypothesis, all you need to do is set your hourglass to wake you up before sunrise. This is when objects off to the other side of the Sun are visible. Within a week or two, your lost planet will emerge. If you repeated these observations, there would soon be no doubt about which planets were coming and which planets were going.

Dear Merlin,

My name is Morgenstern, which translates to "morning star" in English. I assume that the morning star is really the Sun, but I wonder if you can tell me what other stars are seen as the morning star?

DANIEL MORGENSTERN

CLEVELAND HEIGHTS, OHIO

In a logical world, a morning star would indeed be a star.

Traditionally, however, any bright visible planet in the dawn sky of the rising Sun is called a morning star. The best-known morning star is the bright planet Venus, which can rival the landing lights of incoming airplanes for your attention. Since Venus does not actually come in for a landing at the local airport, its appearance often induces people to make phone calls to urban police departments about UFOs with glowing lights that hover. Other morning stars can be Mercury, Mars, Jupiter, and Saturn.

P.S. If you have any German friends named Abendstern, then substitute "evening" for "morning" and rerecite Merlin's answer to them.

Dear Merlin,

In observing the evening sky, I recently noticed that the planet Venus seemed to be located very high above the horizon. I cannot remember Venus being so high, at least during my observational experience of the past twenty or twenty-five years. Please explain the high position of Venus compared with other times when this planet has been the evening's dominant planet.

JAMES R. SHEPPARD, JR.
MESQUITE, TEXAS

If you haven't noticed it in the past twenty-five years, you have missed a few occasions. About every eight years or so, Venus reaches its maximum eastward angle (greatest elongation) from the Sun after sunset in the early spring. The angle the ecliptic makes with the horizon at that time of year, at sunset, and at that time of day puts Venus a whopping 45 degrees above the horizon.

Dear Merlin,

My stepfather is a poet and was recently simultaneously moonstruck and starstruck when he saw the crescent moon close to Venus in the evening sky. In the car on the way to the airport, as he was leaving for China, he asked me how rare that sight was, given that it shows up on the Turkish flag and elsewhere in Arab culture. I quickly estimated for him that Venus is in the evening sky about 25 percent of the time (morning sky 25 percent, behind the Sun 25 percent, between Earth and the Sun 25 percent). And I estimated that the Moon is in a thin waxing crescent phase for about three nights out of about thirty, or 10 percent of the time. The joint probability, then, of seeing the crescent moon in the evening sky with Venus would be about 2.5 percent (0.25 times 0.1), or about ten nights per year, on average, assuming clear skies. Merlin, how good was my estimate?

RICK RAYFIELD
FAYSTON, VERMONT

*M*erlin is duly impressed, as should be your stepfather.

Dear Merlin,

Three years ago we purchased a hot tub, which we set up on a deck outside our house. We use it almost every night and have become avid stargazers. Over the years we have learned to identify many constellations, nebulas, and individual stars.

We live about eight miles from the beginning of the St. Lawrence River in northern New York. One night, at approximately 9:45 P.M., we observed what looked to be a new star near the constellation Böotes. The "star" appeared dull, then increased in brightness, got extremely bright, and then disappeared, all in about five seconds or less. We are assuming it was a meteor that came directly into Earth's atmosphere and burned up. However, no tail was visible, and it never moved from its original position.

Could you please shed some further light on our observation?

MARGE AND RICH FABEND

LAFARGEVILLE, NEW YORK

*M*erlin agrees that hot tubs make excellent stargazing venues.

Meteors can enter Earth's atmosphere at any angle. The most spectacular ones are those that move across your field of view, leaving a brilliant streak of light as they burn. Occasionally a meteor will head straight down your line of sight. When it does, it will appear exactly as you described.

Dear Merlin,

What is the best meteor shower of the year?

BETH ANN SILVERMAN

GRAND RAPIDS, MICHIGAN

The perennial favorite among amateur astronomers is the Perseid meteor shower, seen near the constellation Perseus between August 10 and 14. At a typical rate of one or two meteors per minute, it is among the most rewarding of the year. The best way to watch a meteor shower is to lie still in an open field while looking up for hours and hours after midnight. If the Perseids occurred during the frozen nights of December, then the shower would probably be less popular.

The November Leonid meteor shower, normally undistinguished, can be quite spectacular at the thirty-three-year intervals when Earth passes near the orbiting jetsam of comet Tempel-Tuttle. Merlin remembers back on November 17, 1966, when there were five thousand meteors in a single twenty-minute interval—an average of four per second. Since the meteors radiated from a single location on the sky (the constellation Leo), the scene looked like one of those sci-fi journeys through hyperspace.

Dear Merlin,

> What is a fireball as it relates to meteors?

DARRELL KRUZEN

WISDOM, MONTANA

The term "fireball" is reserved for those blazing, eye-popping meteors that last much longer in the sky than an ordinary shooting star. Fireballs are simply big meteors and are sometimes called bolides.

Some fireballs are bright enough to be seen in the daytime, and many of them are audible as they plow through Earth's atmosphere. Unsuspecting people tend to report fireballs as UFO sightings.

Fireballs are almost always the source of meteor body parts (meteorites) found on Earth's surface.

Dear Merlin,

About twenty-five years ago, in the summertime, I saw some rays of light on the night sky. They came from the north, and they were white in color with ends that came to a point. They even danced around. Could it have been the northern lights?

JEANETTE STEM

KENT, OHIO

$\mathcal{M}$ost decidedly so.

Dear Merlin,

I'll soon visit the Shetland Islands. Will the aurora borealis be visible at night while I am there? Do you have any tips on how to photograph this phenomenon?

ROBERT BAUMGARDNER

AUSTIN, TEXAS

The Shetland Islands of northern Great Britain (latitude 60 degrees north) provide an ideal location to view that cosmic dance of the lights that we call the aurora. The islands, however, are so far north that they get only a few hours of darkness during the nights of the summer months. The aurora (like a shooting star) can happen at any time of day or any day of the year, but is best viewed when the sky is dark. Try to schedule your trip in the late fall, winter, or early spring.

A tripod-mounted video camera will give you a much more rewarding record of the aurora than simple photographs.

P.S. Make sure you have something else to do during your visit. The aurora does not always cooperate with vacation plans.

Dear Merlin,

I recently read a book called Magnetic North: A Trek Across Canada, *by David Halsey. Mr. Halsey claimed that he could hear the aurora borealis, a phenomenon that he stated was rare. Is this true? Can you really hear the aurora borealis?*

BEVERLY SOSNOWSKI

LA GRANGE PARK, ILLINOIS

*A*dd auroras to the growing list of phenomena in the world that are reported to go snap, crackle, and pop.

Dear Merlin,

At about 10:30 P.M. this past July 31, my mom and I saw a small reddish light move quickly across the sky. It tracked through the Big Dipper. It did not blink, like a plane's light might. It was gone from our sight after fifteen seconds or so. Do you think it was a satellite? If so, what kind? And where are they going? How can we see more of them?

CARA AND KERRY DUBYK

FURLONG, PENNSYLVANIA

Since your mother also saw the light, Merlin presumes you were not hallucinating. You probably saw an artificial satellite in low Earth orbit with a polar trajectory. During the several hours after sunset and before sunrise, if a satellite happens to fly over your town, it will be visible because of reflected sunlight before it enters (or after it exits) Earth's cone-shaped shadow in space.

Polar trajectories are ideal for spy satellites because they orbit from pole to pole while Earth turns continuously, which eventually allows the satellite to see every part of Earth's surface, including you and your mother. Next time, wave.

Dear Merlin,

What are magnitudes? In particular, what is the meaning of "apparent magnitude"?

MIKE FRONTZ

GEORGETOWN, TEXAS

*A*stronomical magnitudes are quite simply an inverted, compressed, logarithmic brightness scale with a base equal to the fifth root of one hundred.

The scale is used by most optical astronomers to describe the brightness of stars and galaxies and can be blamed on the Greek astronomer Hipparchus of Nicacea (in about 120 B.C.) and the English astronomer Norman Pogson (in 1856). Each year this magnitude scale confounds tens of thousands of first-year college students who enroll in Introductory Astronomy. You are not alone.

If you *must* know about the magnitude scale, then . . .

The brightness (apparent magnitude) of the star Vega is zero magnitude. Any star that is 2.51188643152... times brighter is assigned magnitude −1. Any star that is 2.51188643152 times brighter still is assigned magnitude −2, and so forth. If negative magnitudes upset you, you will rest easy knowing that there are only nine objects in the sky that are brighter than Vega: Sun, Moon, Mercury, Venus, Mars, Jupiter, and the stars Sirius (the brightest in Canis Major), Canopus (the brightest in Carina), and Alpha Centauri (the brightest in Centaurus). Going the other way yields magnitude +1 for a star that is 2.51188643152 times dimmer than Vega, magnitude +2 for a star that is 2.51188643152 times dimmer than that, and so forth.

For every five magnitudes you get a factor of 2.51188643152 × 2.51188643152 × 2.51188643152 × 2.51188643152 × 2.51188643152 in brightness. People who never leave home without their calculator will notice immediately that five magnitudes represent a factor of one hundred in brightness. (This defines a logarithmic measuring scale with a base equal to the fifth root of one hundred.)

The brightness (apparent magnitudes) of some familiar objects on the magnitude scale:

Sun	−26.7
full Moon	−12.6
Venus at its brightest	− 4.4
Sirius	− 1.5
North Star (Polaris)	+ 2.0
Merlin's home (Andromeda galaxy)	+ 4.4
limit of unaided human eye	+ 6
Barnard's star	+ 9.6
quasar 3C273	+12.9
detection limit of 200-inch Mount Palomar telescope	+24

Note that nearly fifty-one magnitudes separate the brightness of the Sun and the detection limit of Mount Palomar's telescope—a factor of 200 quintillion in brightness.

You may occasionally see the term "absolute magnitude," which is defined as the apparent magnitude an object would have if it were placed at a distance of 32.6 light-years (10 parsecs). It forms a comparative measure of light emitted. The absolute magnitudes of our familiar objects become:

quasar 3C273	−26.1
Merlin's home	
(Andromeda galaxy)	−19.6
North Star (Polaris)	− 4.5
Sirius	+ 1.4
Sun	+ 4.7
Barnard's star	+13.2
Venus at its brightest	+27.2
full Moon	+30.4

In absolute magnitudes the quasar wins, the Sun is undistinguished, and the full Moon loses badly.

Dear Merlin,

I have a hard time accepting the notion that a star can have zero magnitude yet be one of the brightest objects in the sky.

ELMIRA JENKINS

KANSAS CITY, MISSOURI

$\mathcal{M}$erlin bets that you have accepted other meaningful zeroes in your life. Do you ever ask weather forecasters, "How can it be zero degrees outside?" Probably not, because you know (or should know) that zero degrees Celsius is just someplace on a temperature scale (the freezing point of water) and that the entire scale innocently includes positive and negative numbers.

Similarly, if you have zero dollars, that's a whole lot more money than if you owed somebody a thousand.

Dear Merlin,

Astronomers use the term "right ascension," measured in hours, to describe one coordinate for locating bodies on the celestial sphere. But why? Nowhere have I been able to learn the authoritative etiology of the term "right ascension." It is no doubt very old, and has been used in astronomy and celestial navigation for many years. Why do we continue to use it when a term like "celestial longitude" would be more appropriate, perhaps? Your omniscient assistance would be appreciated.

CARL F. BACHLE

JACKSON, MICHIGAN

The term "right ascension" was invented by the Greek astonomer Hipparchus. Merlin recently chatted with the Polish astronomer Nicolaus Copernicus about stars, the planets, and Hipparchus's celestial coordinate system. The year was 1540. Nicolaus and Merlin were in a meadow near the town of Frauenberg, East Prussia.

MERLIN: Nick, why is "right ascension" such a funny name?

COPERNICUS: It was not intended to make people laugh.

MERLIN: ?!

COPERNICUS: It is all quite simple. Following the ideas of the Greek astronomer Hipparchus, the sphere that encloses an observer where stars rise and set at right angles to the horizon is called a "right sphere."

MERLIN: This is true, of course, for observers on earth's equator.

COPERNICUS: Yes. But for people anywhere else on earth (other than the poles), stars will rise oblique to the horizon. These observers are enclosed by "oblique spheres."

MERLIN: How does this relate to right ascension?

COPERNICUS: As the earth rotates, one can measure a star's ascension above the horizon on the oblique sphere. But one can also measure a star's ascension on the equator's right sphere. Merlin, it should be clear that a star's "right ascension" is a much more useful coordinate than a star's "oblique ascension"—especially since different latitudes on earth see different oblique ascensions.

MERLIN: Have you told others about Hipparchus's coordinate system?

COPERNICUS: It is fully documented in Book II of my much larger work, where I find that the sun, not earth, is likely to be the center of planetary motion. I call it *De revolutionibus orbium coelestium*, but I have not yet published it.

MERLIN: You'd better hurry, you are already sixty-seven years old. You haven't much time to waste.

Copernicus died three years later. *On the Revolutions of Heavenly Bodies* was published while he lay on his deathbed.

Dear Merlin,

I understand that declination in the sky corresponds to latitude on Earth while right ascension in the sky corresponds to longitude on Earth. But if longitude starts at an imaginary line running through Greenwich, England, then where does right ascension begin? I know it is supposed to be where the Sun crosses the celestial equator (about March 21), but where is the actual point in the sky, and how can I look at stars and roughly speculate on their right ascension?

KEVIN ORLOWSKI

TEMPE, ARIZONA

The point in the sky on the celestial equator where right ascension (RA) equals zero hours and zero minutes is near the boundary between the constellations Pisces and Aquarius. This spot may also be found 15 degrees south of the more easily recognized "great square" in the constellation Pegasus.

You can estimate right ascension by aiming your right arm 15 degrees south of Pegasus and spanning your left arm eastward. The celestial equator also passes near the three bright stars in the belt of the constellation Orion. If you make sure the arc you trace passes through Orion, the angle between your two arms immediately indicates right ascension. Half a right angle is 3 hours RA. A right angle is 6 hours RA. Arms spread wide is 12 hours RA. (Traffic police and school crossing guards tend to be good at this.) If you plant your left arm 15 degrees south of Pegasus and span your right arm westward, you subtract the hours you get from 24. For half a right angle you are at 21 hours RA. For a right angle you are at 18 hours RA.

This procedure will help you estimate the RA along the

celestial equator. For the rest of the sky, note that stars in line between the north celestial pole and the south celestial pole through a chosen spot on the celestial equator all have the same RA.

Of course, all this arm-swinging in the dark can be avoided if you have at your disposal a detailed star chart from any one of the monthly amateur astronomy magazines.

Dear Merlin,

I would like to know what these "M" designations are in globular cluster names, for example M15 or M54. Are they just numbers given to these objects by figuring their distance?

RENEE TINSLEY

STOCKTON, CALIFORNIA

*I*t was the year 1771 when the French astronomer Charles Messier published his first list of fuzzy-looking objects in the sky (stars, clusters, and nebulas). His intent was to provide a false-alarm list for comet hunters.

Just over one hundred objects have his numerical classification. Some famous ones are Messier 31 (the Andromeda galaxy), Messier 13 (the Hercules globular cluster), Messier 42 (the Orion Nebula), and Messier 1 (the Crab Nebula). Some less-than-famous objects are Messier 102 (an accidental repeat of Messier 101) and Messier 91 (it does not exist).

For brevity, the *essier* is often dropped, so that one simply writes M before the number.

Dear Merlin,

I have a question about information in two versions of my astronomical field guide. One was printed in 1964 and the other in 1983. The information about specific stars conflicts between the two editions. For example, in the 1983 guide it says that the star Deneb in the constellation Cygnus is 1,830 light-years away. In the older guide it says that Deneb is 600 light-years away. I have also found many other differences. Why are they so different?

JEFF DETRICK

ARNOLD, CALIFORNIA

*A*s in all sciences, experimental or observational results have ranges of uncertainty. When scientific papers are published in journals, these uncertainties are reported. When books for the general population are written, these uncertainties are almost always omitted.

Consequently, when you see a size or distance listed in a book, you usually have no idea how well the quantity is known. You should not be alarmed that in the twenty years that span the two editions of your book, some entities of the universe have become known a little bit better.

Dear Merlin,

My question regards the constellation Cancer, the Crab. Where is it? I don't often see reference to it on star maps. I am curious, of course, because it is my astrological sign. However, astronomy, not astrology, is primary and brings forth the question.

SHIRLEY HARTWELL

GUILFORD, CONNECTICUT

The brightest four stars in Cancer are all relatively dim. With extraordinary imagination and an abandonment of common sense, you can see a two-clawed crustacean in the sky. Look between Leo and Gemini.

Please remember: astronomers seek to understand the stars; astrologers seek to be ruled by them.

Dear Merlin,

Is Orion, the constellation, right- or left-handed? I have seen references to the star Betelgeuse as being his left shoulder. Is Orion facing toward or away from us? I have always envisioned him as facing toward me and being right-handed, in which case Betelgeuse is his right shoulder.

LT. COL. ROBERT A. FOWLER
SAN ANTONIO, TEXAS

$\mathcal{M}$erlin once chatted with Homer as he was composing the *Iliad* back in the ninth century B.C. He said of the famed hunter, "Illustrious Orion, the tallest and most beautiful of men . . ." But of course Homer was blind. Early depictions of Orion in the sky have his back to the viewer, with his face in profile as he wisely takes note of Taurus the bull over his right shoulder. This makes him a lefty, since his left hand wields the weapon while his right hand holds the arc of stars that compose his shield.

Recent constellation artists, however (probably righties), have turned him around to afford a better view of his beauty—although between the two bright stars that are his shoulders and above the three dim stars that compose his neck there is a big empty region where Orion's head ought to be. So feel free to imagine Orion as you please.

VII

THE WORLD
of
LIGHT

Dear Merlin,

Can you see starlight from the bottom of a well in the daytime?
FREIDA HAWTHORNE
SEATTLE, WASHINGTON

*N*o. All the people who said you could have simply never climbed to the bottom of a well to check—possibly because they didn't have a well to jump into. And those who do own wells probably would not want to jump into them anyway.

Earth's daytime atmosphere is aglow with scattered "sky-blue" light from the Sun. Starlight is summarily swamped in the presence of this glow. Since the atmosphere glows no less when viewed from the bottom of a well, what you really need to see the stars in the daytime is a hole through the atmosphere, not a hole in the ground.

Dear Merlin,

If photosynthesis is the formation of carbohydrates in the chlorophyll-containing tissues of plants exposed to light energy, then it seems to follow that energy, to some degree, is being transformed into mass and thus matter is being created. If this is true, a lot of people are wrong. Plus, Earth should weigh more now than it did before plants appeared. Is this true? And if it isn't, would you please straighten me out?

TONY GRANADOS

BENSALEM, PENNSYLVANIA

There are many ways that energy can be converted into matter but photosynthesis is not one of them.

Acorns, that tasty food of squirrels and many other rodents with big front teeth, grow into ten-ton oak trees because of what they absorb from the ground and from the air. Sunlight allows a tree to make wood out of atmospheric gases, water, and minerals. If you plant a tree, you will help to remove carbon dioxide from Earth's atmosphere.

Let us not forget that the atmosphere of Earth's sibling planet, Venus, is mostly carbon dioxide, which has induced a greenhouse effect that sustains a hotter-than-a-pizza-oven surface temperature of 900 degrees Fahrenheit.

Dear Merlin,

Who was the first person to invent the telescope?

SELENA RAMM

CLAREMORE, OKLAHOMA

It was 1608 when the Dutch spectacle-maker Hans Lippershey put two lenses in a line to produce the first telescope. At seven power, the Lippershey telescope was more a curiosity than a useful scientific instrument.

Two years later, Galileo developed a telescope that was more advanced than Lippershey's original. One spring afternoon in Venice, Galileo told Merlin, "By sparing neither labor nor expense, I constructed for myself an instrument so superior that objects seen through it appear magnified nearly a thousand times, and more than thirty times nearer than if viewed by the natural powers of sight alone."

Galileo originally called his improved telescope a spyglass, which makes Merlin wonder which heavenly bodies he might have looked at first. Galileo eventually discovered lunar craters and mountains, sunspots, the phases of Venus, Jupiter's four largest moons, Saturn's rings, and the stars of the Milky Way.

Dear Merlin,

How do telescopes work? How is their power determined?

ABE SHULTZ

DELMAR, MARYLAND

Telescopes of all varieties have a simple mission: to collect and focus light. Yes, some telescopes focus better than others, and some telescopes are larger than others, but they all serve to extend the breadth and depth of human vision into the universe.

The most powerful telescopes are normally the largest. They collect the most light and allow the observer to see the dimmest objects. Light-collecting power grows as the square of the telescope size. For example, a sixty-inch telescope has three times the diameter of a twenty-inch telescope, but it has nine times ($3^2 = 3 \times 3 = 9$) the surface area and thus nine times the light-collecting power. Another important advantage of a large telescope is the improved ability to resolve detail.

Further aspects of a telescope's design are influenced by the type of light that the telescope will observe. Gamma-ray telescopes, X-ray telescopes, optical telescopes, microwave telescopes, and radio telescopes are all built differently. For example, the design and materials that can focus radio waves (such as a large concave dish made of chicken wire) will not focus optical light, and the design and materials that focus optical light (such as polished and aluminized glass) will not focus X-rays.

To say that telescopes magnify is misleading. On amateur optical telescopes, the magnification is set by the choice of eyepiece, not by the telescope. Note that the ability to detect a dim object will not be improved by an eyepiece with higher magnification. And if the telescope itself makes bad images, then all you will do with your high-powered eyepiece is magnify a bad image.

A complete telescope design requires a detector placed where the light comes to focus. In the days of Galileo and Newton, biological detectors such as the retina of the human eyeball were common. The 1800s brought forth chemical detectors called photographic plates. Nowadays, digital detectors reign supreme.

Dear Merlin,

I am trying to put in terrestrial perspective the light-gathering ability and magnifying power of large optical telescopes. If the five-meter telescope at Mount Palomar was a telephoto lens for my 35mm camera, what would be the focal length and maximum focal ratio?

JOHN J. O'DONNELL

VIRGINIA BEACH, VIRGINIA

*I*f you attached the Mount Palomar five-meter telescope to your handheld camera, you could take awesome photos. The maximum (brightest) focal ratio would be ƒ/3.25, with a focal length of 16,460 millimeters. By reattaching the telescope in different ways, you could achieve a focal length of 149,960 millimeters, with a focal ratio of ƒ/30.

For convenience, however, we normally attach cameras to telescopes rather than telescopes to cameras.

Dear Merlin,

How many photons do we have to gather from a distant light source to make a spectrum? Is one enough?

MARY LEE COLEMAN

PARSIPPANY, NEW JERSEY

Conceptually, a single photon carries information only about a single part of the spectrum. A million such photons would be a rather boring spectrum, because they would all be the same color.

Experimentally, a photon must trigger a detector for us to be aware of its existence. In humans, the detector is the retina. In modern telescopes, the preferred detector depends on the part of the spectrum being observed.

The sensitivity of modern detectors is determined by how many photons it takes to trigger a measurable electrical signal. Only one or two photons are required to trigger today's best visible light detectors—the silicon-based charged coupled devices (CCDs) are thousands of times more sensitive than the earliest photographic detectors.

The duty of a prism or diffraction grating is to spread a mixed collection of photons in such a way that a smooth and continuous spectrum is produced. When the spectrum leaves its image on a CCD, the photon energies become grouped into discrete columns across the detector. If, for example, a CCD has 2,048 columns, then to trigger *and* span the detector you need at least 2,048 photons. The most reliable spectra, however, are obtained when many more than the bare minimum of photons are received.

Dear Merlin,

How was the speed of light first determined?

LOREN FERCH

NEW HOPE, MINNESOTA

One of the first attempts to measure the speed of light was made by Merlin's good friend Galileo Galilei in the early 1600s. He sent an assistant to a distant hill to flash the light of a lantern. Galileo responded as fast as he could with flashes from a lantern of his own. His attempt to time the delay proved futile—human reflexes were inadequate for such a task. Afterward, Merlin overheard Galileo comment that light, "if not instantaneous . . . is extraordinarily rapid."

In 1675 the Danish astronomer Ole Rømer noticed that eclipses of Jupiter's moons occurred earlier than expected when Earth was nearest Jupiter and occurred later than expected when Earth was farthest from Jupiter. Rømer correctly blamed these differences on the time it must take for light to span the diameter of Earth's orbit. By making a simple division— diameter of Earth orbit divided by total time difference equals speed of light—Rømer provided the first reasonable estimate for the speed of light.

Nowadays, the speed of light can be measured easily with tabletop interferometers, which are fancy devices that use light's own optical properties to make high-precision measurements.

In case you are wondering, the modern value for the speed of light is exactly 299,792,458 meters per second.

Dear Merlin,

Why is it impossible ever to attain the speed of light?
DANIEL L. RHONE
BEVERLY, NEW JERSEY

There are countless "why" questions that do not lend themselves to scientific investigation. Often such questions probe the intent of the universe, like "Why do atoms exist?" or "Why was there a big bang?" Your question is best reworded, "What happens if you try to attain the speed of light?"

As Merlin's good friend Albert Einstein described in his special theory of relativity, if you approach the speed of light in your spaceship, an outside observer will notice that your mass will increase, your time will slow down, and your spaceship's length (and all its contents) will shorten. Einstein noticed from his equations that at the speed of light, mass is infinite, time stops, and length is zero. He wisely concluded that it must be impossible to travel at the speed of light.

All experiments ever conducted (mostly with fast-moving subatomic particles) have confirmed these predictions.

Dear Merlin,
    What was the Michelson-Morley ether experiment?
    MARY LEE COLEMAN
    PARSIPPANY, NEW JERSEY

The American physicist Albert Michelson and the chemist Edward Morley collaborated in 1887 on a project to measure the speed of light in the same direction as Earth in its orbit and at right angles to it. This led to one of the most famous no-result experiments in all of physics.

Michelson and Morley attempted to show that space was filled with a hypothetical substance called ether that permitted waves of light to travel in much the same way as air permits sound waves to travel. If the ether existed, the measured speed of light in the direction of Earth's motion would be different from the speed determined at right angles to Earth's motion.

There was no variation in the speed of light.

The experiment simultaneously killed the long-held concept of the ether and showed that the speed of light is independent of the motion of the observer, thus paving the way for the founding principles of Einstein's special theory of relativity in 1905.

Dear Merlin,

Recently a creationist on a radio show stated that the speed of light has not always been constant, probably slowing down, thus accounting for biblical time references. This does not sound plausible, but would you please comment?

CHUCK MILLER

SHAWNEE, KANSAS

There is no evidence to suggest that the speed of light has ever been anything other than constant for all of space and for all of time.

Dear Merlin,

Has anyone experimentally measured the velocity of light originating beyond the solar system? Beyond the Milky Way galaxy?

AUBREY R. MCKINNEY, PH.D.

CORPUS CHRISTI, TEXAS

The English astronomer James Bradley first measured in 1725 what he later called the "aberration of starlight," which simply meant that his telescope had to be angled slightly (20½ arc seconds) from the expected position of the target star. Bradley deduced correctly that the angle depends on the speed of Earth in its orbit and on the speed of light. This aberration angle is conceptually no different from the angle that a vertically falling raindrop makes as it streaks diagonally down your car door's window. If your window had little adjustable buckets hinged to the glass, you would have to tip them to the angle of the streaks if you wanted the rain to go "straight" into the buckets. Fatter raindrops that fell faster than average would streak at an angle that was closer to the vertical. Indeed (for a given car speed), the raindrop angle is a direct measure of a raindrop's speed.

If the light that originated from quasars came to Earth with a different speed from the light that originated in nearby galaxies or stars or planets, each object would display a different aberration angle. But they don't.

Dear Merlin,

Do you or any of your cousins out there know of a way around the limit imposed by the speed of light? If so, please expound.

S. J. TEMPLETON

CHICAGO, ILLINOIS

No.

The speed of light is not just a good idea. It's the law.

Dear Merlin,

After all the problems we have with government-imposed speed limits, why did freethinking Albert Einstein impose one on the universe with the speed of light, yet Isaac Newton did not?

S. J. TEMPLETON

CHICAGO, ILLINOIS

You cannot blame scientists for discoveries of the universe that you do not like. It would be like blaming the mailman for the contents of your mail. And unlike governments that meddle in their citizens' affairs, scientists do not tell the universe what to do or how to do it. Albert simply recognized a preexisting speed limit where no one else had thought there was one. Therein is further, not less, evidence of a freethinking mind.

Dear Merlin,

    How much light pressure will a strong laser put out?

FRANCIS BLOCK

ROBSTOWN, TEXAS

The strongest lasers available concentrate light into small, short pulses. Typically, they will put 10^{12} (one trillion) watts of light power into the size of a pinhead for one billionth of a second.

Americans commonly think of pressure in pounds per square inch. Since these laser beams are narrow, one may prefer to think of pounds per pinhead. Note, however, that laser targets usually convert the laser power into heat rather than motion.

Provided the laser beam does not vaporize your target, a pressure of about one-quarter pound per pinhead can be imparted with each pulse.

Dear Merlin,
   Could laser light be used to propel a spaceship to the stars?
   FRANCIS M. BLOCK
   ROBSTOWN, TEXAS

A laser rocket, or more generally a photon rocket, is the most efficient means of propulsion known. In other words, compared with standard chemical fuels, a photon rocket has more bounce to the ounce. The spaceship engines would systematically need to convert mass into photons (the Sun does this to over 100 million tons of matter each second) and aim the photons out the rear. In response, the ship would recoil forward.

   The problem is that if you wish to travel to a star fifty light-years away (relatively close on the scale of the Milky Way galaxy), and if you wish to arrive before you die of old age, nearly 90 percent of the spaceship's mass would have to be converted into photons for you to get up enough speed. Unfortunately, you would need another 90 percent of the spaceship's mass to slow down again to avoid a crash landing. If you expect to return to Earth, you must forfeit yet another 90 percent of the spaceship to speed up and then another 90 percent to slow down again for your triumphant return.

   If you and your life support weigh 2,000 pounds, then all this 90-percent-of-the-spaceship mass loss would require your initial rocket to weigh 10,000 tons (equivalent to about twenty Boeing 747s). As hopeless as this scenario looks, any other known form of propulsion is worse.

Dear Merlin,

　Assume an electron is captured by an atom directly into its lowest energy state and all its energy is dumped into a single photon. The resulting photon is capable of ionizing the next atom it encounters. Is it correct to think of a massless photon packet as a concentrated bundle traveling at the speed of light, in the same manner as a projectile fired from a gun?

　RON GIFFIN
　CINCINNATI, OHIO

The same laws of physics that describe the recoil of a gun also apply to the atom that emits a photon. Photons and projectiles each have energy and momentum that they carry away. The only important differences between them is that the photon does not get its energy from gunpowder—and it will not accidentally kill a member of your family.

Dear Merlin,

My husband told me that he heard a news report that astronomers have found something that travels faster than the speed of light. Is this true?

IRIS LEWIS
DEBARY, FLORIDA

No.

Dear Merlin,

To my understanding, Einstein's theory says that anything that has mass cannot go the speed of light. If light can be bent and even caught by gravity (black holes), it would indicate to me that it has mass. How can light have mass and still go the speed of light?

RICK LOWE

AURORA, COLORADO

$\mathcal{D}$on't think of light bending because it has mass (which it doesn't); think of light bending because space bends. According to Einstein's general theory of relativity, first published in 1916, gravity is the curvature of space.

Armed with this simple yet revolutionary concept, one can easily see that massless light merely follows the fabric of space, curved or otherwise.

Dear Merlin,

It is stated by Einstein's special theory of relativity that nothing can travel faster than the speed of light. Then how is it that Cherenkov radiation can be light emitted by a high-speed charged particle that passes through material at a speed greater than the speed of light? How is this possible?

ROGER POWERS

CHELAN, WASHINGTON

The speed of light through the vacuum of space is 186,282 miles per second. Nothing has ever been measured to travel faster than the speed of light, and nothing ever will. This is one of the two principles of the special theory of relativity.

When Einstein came up with this theory, little did the Soviet physicist Pavel Cherenkov know (he was just one year old) that he would later discover that radiation is emitted when a charged particle induces an electromagnetic "shock wave" as it moves faster than light in a transparent medium (water, glass, etc.). No speeding tickets here, however, because the speed of light through matter can be much less than the vacuum speed of 186,282 miles per second. In water, for example, the speed of visible light is about 75 percent of that value; through a diamond, a mere 40 percent. A properly accelerated subatomic particle will have no trouble passing these speeds and consequently emitting Cherenkov radiation.

Dear Merlin,

I was having dinner in a famous Santa Barbara restaurant last night and overheard part of a conversation between two gentlemen at another table about the speed of gravity being greater than the speed of light! I never heard of the speed of gravity. Is there a speed for gravity? Is it faster than the speed of light?

MIKE GRANT
SANTA BARBARA, CALIFORNIA

*F*ood for thought: changes in gravity exert themselves through "carrier" particles called gravitons that travel at exactly the speed of light as predicted in Albert Einstein's general theory of relativity.

These much-dined-over particles have yet to be detected.

Dear Merlin,

    Doesn't a black hole have less mass than the star from which it originates? Then why (in artistic renditions) are black holes frequently shown to be gobbling up their binary companion's atmosphere?

CYNTHIA FRAY

LONG BEACH, MISSISSIPPI

Yes. Black holes that are born from the death of high-mass stars typically have only a fraction of the mass of the original star. But the artists are not lying to you.

    The black hole cannot dine upon the companion star until the companion star swells to become a red giant. If the two stars are close enough to each other, then the red giant will lose its gravitational grip on its tenuous atmosphere as its outer layers spiral toward the black hole. When we "detect" black holes, we measure the ultraviolet light and X-rays emitted by this funneling gas.

    At any previous moment, an artist's rendition would be uninteresting.

VIII

THE WORLD of
PHYSICAL LAWS

Dear Merlin,

What is the origin of the Fahrenheit temperature scale? In particular, please explain the temperatures assigned to the freezing and boiling points of water (32 degrees and 212 degrees). They have always been a mystery to me.

R*OBERT* L*UPTON*

R*UGBY,* E*NGLAND*

$\mathcal{M}$erlin happened to meet the German physicist Gabriel Daniel Fahrenheit in an Amsterdam tulip garden during the early spring of 1729, while Fahrenheit was amid a lengthy stay in the Netherlands. Merlin and Fahrenheit chatted about thermometers while enjoying a snack of herring and genever.

MERLIN: Gabriel, I hear from many people that you have constructed some impressive thermometers.

FAHRENHEIT: Yes, Merlin. I have worked for quite some time on a calibrated and trustworthy thermometer. It contains quicksilver as the liquid indicator rather than alcohol.

MERLIN: But what is the origin of your mysterious temperature scale?

FAHRENHEIT: The idea is loosely based on the zero-to-60-degree scale used by the Danish astronomer Ole Rømer in his studies of the weather. Rømer declared 60 degrees to be the boiling point of water, and then, to insure that he would not greet negative temperatures, he placed one eighth of the scale below the freezing point of water. On Rømer's scale, a mix of equal parts of water and sea salt coincidentally freezes near zero, but ordinary water freezes at 7½ degrees, while water that is lukewarm to the touch is 22½ degrees.

MERLIN: What changes did you make to this scale?

FAHRENHEIT: I further subdivided each of Rømer's degrees into four equal parts. With this change, the temperature of freezing water becomes 30 degrees, and that of lukewarm water becomes 90 degrees. I then decided to use lukewarm water as a reference rather than boiling water because of the readiness with which it can be obtained. It is simply the blood temperature of a healthy adult, which can be measured easily from the mouth or from the pit of the arm.

MERLIN: Armpits are not likely to be tasty. It may be a good idea to keep at least two *separate* thermometers in the home . . .

FAHRENHEIT: So I have discovered. As I was saying, with blood temperature as my reference, I then rescaled the zero-to-90 degrees into a 96-degree interval. With this adjustment, water is now measured to freeze at 32 degrees and the new interval is evenly divisible by many numbers: 1, 2, 3, 4, 6, 8, 12, 16, 24, 32, and 48. It is now trivial to subdivide each degree into fractions when it comes time to etch the brass holder-plate for the thermometer.

MERLIN: How 'bout that 212?

FAHRENHEIT: When I carefully correct for the expansion of the thin glass tube that contains the quicksilver, I find that rainwater boils at 205½ degrees, while other waters boil as high as 212½ degrees. I also find that all types of water will boil at lower temperatures when moved to high altitudes. Of course, I had to leave the Netherlands to discover this.

MERLIN: Congratulations on your efforts. Keep up the good work. Perhaps one day your temperature scale will be officially adopted. More herring?

In 1777, the Royal Society of London, in a committee headed by the esteemed British physicist Henry Cavendish, defined the sea-level boiling point of pure water on the widely used Fahrenheit scale to be exactly 212 degrees, and the Fahrenheit scale was officially sanctioned. This nudged the human body temperature upward from 96 degrees F to the commonly quoted value of 98.6 degrees F. The freezing point of water remained exactly 32 degrees F.

Unfortunately for Fahrenheit, his fraction-friendly scale was ultimately replaced by the decimal-based Celsius scale among the world's scientists and in nearly every civilized nation on Earth.

Dear Merlin,
What is a particle accelerator and how does it work?
SHORTI THOMPSON
BEDFORD, VIRGINIA

The job of a particle accelerator is to accelerate particles.

Charged particles accelerate in a predictable way in the presence of magnetic or electric fields. This is how negatively charged electrons know where to go before they strike the phosphors on the inside of your TV screen. The field strength and electron energies within your TV set, however, are not high enough to smash atoms.

In a specially designed cavity (it can be circular or straight) you can rig a changing magnetic and electric field in such a way that particles move faster and faster until they have sufficient energy to smash your target atoms or, more generally, to explore unknown nuclear interactions. The longer the path and the stronger the field, the faster the particles can move and the more exotic is your particle-smashing experiment.

With high enough energy, the conditions within the particle accelerator begin to resemble the cosmic primordial soup, where a shotgun marriage has taken place between physicists who study subatomic particles and astrophysicists who study the prevailing conditions of the early universe.

Dear Merlin,

How many elements are there? I can only name a dozen or so, but I'm sure there must be more.

JEREMY BLUSTEIN

BROOKLYN, NEW YORK

Your dozen probably includes some of the most commonly discussed elements: hydrogen, helium, carbon, oxygen, sulfur, chlorine, neon, iron, gold, silver, and platinum. A hundred more elements bring the grand total to above 110.

Each element has a unique "atomic number" that identifies the number of protons in its nucleus. At atomic number 1 we have the smallest element, hydrogen, with one proton in its nucleus. At atomic number 92 we find the largest naturally occurring element, uranium, with ninety-two protons packed in its nucleus.

Dear Merlin,

Why was the element cesium chosen to measure time?

G. K. STANTON

NORMAN, OKLAHOMA

For those who missed it, on October 13, 1967, at the Thirteenth General Conference of Weights and Measures, the second was defined as "the duration of 9,192,631,770 periods of the radiation corresponding to the transition between two hyperfine levels of the fundamental state of the atom of cesium 133."

If you were inclined to build an atomic clock of your own, and if you are especially punctual with your engagements, you would have been ecstatic. Cesium's chosen atomic transition allows a precision of measurement down to one part in 100 billion, which adds up to about a third of a second every thousand years.

Dear Merlin,

Would two magnets attracting or repelling each other eventually lose their energy and lose their magnetism in the process?

JOSHUA SHINAVIER

SPOKANE, WASHINGTON

It depends on whether you are nice to your magnets. If two magnets are forcibly strapped together with like poles touching, they will eventually demagnetize each other. If the magnets are slammed on the ground or heated, they will also become demagnetized.

But if your magnets are minding their own business, or if they are attached to each other with opposite poles touching, they will stay magnetized forever. In other words, it takes energy to demagnetize a magnet.

Dear Merlin,

I always thought water was incompressible. How can it be compressed in a black hole?

G. R. JOHNSON
PHOENIX, ARIZONA

"Water is incompressible" is one of the great lies of society, on a par with "What goes up must come down." Each of these is true for most experiments that you can perform, but neither describes the physical limits of the universe, as does "Nothing can travel faster than light in a vacuum."

It is not widely appreciated that atoms contain mostly empty space. If you removed this empty space from all the water in Earth's Atlantic and Pacific Oceans, together they would fill a cube about fifteen yards on a side.

Black holes can be easily manufactured in the high-pressure core of a collapsing high-mass star, but until cookbooks have recipes for homemade black holes, the "Water is incompressible" lie will remain with us.

Dear Merlin,

When we heat water, the molecules start to move faster and form a gas that we know as steam. If oxygen is a gas, what would happen if we could possibly slow down the molecules of the oxygen? Would it be possible to form oxygen into a liquid or a solid?

GINA YOSTEN (AGE 12)

MUENSTER, TEXAS

Yes. If you chilled air down to about −300 degrees Fahrenheit, you would liquefy its oxygen content. And if liquid oxygen does not impress you, then drop the temperature another 60 degrees to freeze it solid.

Merlin recommends that you get your parents' permission before you try this experiment at home.

Dear Merlin,

If $E = mc^2$, then why does all the literature state that the limitation is the speed of light and not the speed of light squared?

D<small>AVE</small> K<small>ISOR</small>

P<small>HOENIX</small>, A<small>RIZONA</small>

*W*ith his famous equation, published as part of his special theory of relativity, Albert Einstein simply showed that mass and energy are equivalent. To convert the units of mass into the units of the energy, one needs a constant, which in this case happens to be the speed of light squared. There are even some physics equations where the speed of light is raised to the third and fourth powers. But none of this has anything to do with speed limits.

Dear Merlin,

Can Einstein's famous equation $E = mc^2$ be derived from more basic forms?

MICHAEL LAWLOR

SALISBURY, NORTH CAROLINA

There is no equation more fundamental than Albert Einstein's $E = mc^2$. It provides the recipe to compute the physical equivalence of an object's mass at rest and its energy. Sir Isaac Newton's $F = ma$ is another equation that cannot be derived from more fundamental forms. It provides the recipe to compute the acceleration induced by a force on a mass.

Note, however, that the existence of these equations can be reasoned from assorted secondary arguments, which is often done in textbooks.

Dear Merlin,

I wonder if it is possible to create a string of mathematical identities to go from $E = mc^2$ to $F = ma$. Perhaps a computer loaded with mathematical equations could be programmed to do it, somewhat in the manner of trigonometric equations.

PAUL EVANS
PLANO, TEXAS

*N*o, it is not possible.

In pure mathematics, the job of the equals sign is to connect two quantities that are numerically equal. In the real world, however, the equals sign connects two quantities that are numerically equal and physically equal. Albert Einstein's equation measures (E)nergy. Isaac Newton's equation measures (F)orce. You could adjust the arithmetic so that the numerical value of *F* equals that of *E*, but if you wrote $E = F$ it would have positively no physical meaning. Does a 55-cent can of tomato paste equal the 55mph speed limit on the highways?

Dear Merlin,

Currently I am studying various ways of proving or demonstrating that Earth rotates. The Foucault pendulum concept is confusing to me. Will you please explain why the Foucault's pendulum experiment proves that Earth rotates?

James O. Roberson

Williamston, North Carolina

The French physicist Jean Bernard Léon Foucault first demonstrated in 1851 the pendulum that displayed Earth's rotation. Merlin was present at this demonstration, where the following conversation unfolded.

MERLIN: Jean, pendulums have been around since the days of Galileo. What have you contributed to our knowledge that is not already known? They say you can prove that Earth rotates.

JEAN: Good to see you, Merlin. If I make the wire of a pendulum long and thin enough, and if I make the weight at the bottom heavy enough, then the pendulum can swing for many days before it comes to rest from the friction at the pivot.

MERLIN: How can this demonstrate the rotation of Earth?

JEAN: It is quite simple. If you place such a pendulum directly over Earth's North Pole and set it to swing, then Earth will rotate beneath it. As you stand on Earth, it will look as if the pendulum slowly changes its direction of swing over the 23-hour-and-56-minute rotation period of Earth.

MERLIN: That is remarkable—you should give one of your pendulums to Santa as a present for Christmas. Will this also work on the equator?

JEAN: No, Merlin. As you move southward from the North Pole, the effect of the rotating Earth becomes less and less. With some complicated mathematics, I can show that the pendulum responds by taking longer and longer for its direction of swing to make a full circle. In Paris—latitude 49 degrees north—my pendulum takes nearly 32 hours. In Dakar—latitude 15 degrees north—the period is nearly 93 hours. In Singapore—latitude 2 degrees north—people would be quite bored with my pendulum; a full cycle there would take over two years. And on the equator, of course, the swing of the pendulum remains fixed.

A year later, Jean Foucault invented the gyroscope, although today he is best known for his pendulum.

Dear Merlin,

With today's technology in determining solar system orbits, have we succeeded in improving the three-digit accuracy of the gravitational constant since my textbook days in college during the mid-1950s?

RON GIFFIN

MONTGOMERY, OHIO

*N*ewton's gravitational constant is often reported to be 6.67×10^{-11} m^3/kg/s^2, but we can give you several more decimal places if you prefer: 6.67259×10^{-11} m^3/kg/s^2. This is the best we can do for you now, but there remains some uncertainty in the last two digits.

The computation of orbits is a complicated business. The major uncertainties are, and have always been, the proper accounting for all gravitational sources, such as planets, moons, and the major asteroids, and their mutual interactions over time.

Dear Merlin,
    Does the Sun's gravity affect our weight on Earth? Do we weigh more at midnight than we do at noon?
    LESTER HELLMAN
    PRESCOTT VALLEY, ARIZONA

If you feel lightheaded under the midday Arizona sun, please do not blame it on the Sun's gravity.

The Moon orbits Earth. While the *Apollo* astronauts bounced around on the Moon, Earth's gravity had no effect on their weight. Similarly, while we stand on Earth's surface, we (and Earth) are in orbit around the Sun. As a result, the Sun's gravity has no effect on our weight.

These facts are a general property of gravitational orbits, because orbits are dynamically equivalent to free fall. Merlin's good friend Sir Isaac Newton first reasoned this back in 1686.

Dear Merlin,

The Shoemaker-Levy 9 comet encounter with Jupiter back in July 1994 has me thinking. On its previous pass by Jupiter, a single comet broke apart into many pieces. What is the magnitude of the forces that are responsible for this? On Earth these forces give rise to the ocean tides. Earth's 8,000-mile diameter makes a force that sustains a column of water five to ten feet high. If comet diameters are about a factor of 1,000 less than Earth's, then can it be possible that the tidal forces on the comet are anywhere near that of Earth's tidal force?

RON GIFFIN

MONTGOMERY, OHIO

Tides are simply the result of the difference in gravitational force from one side of an object to the other as caused by some nearby source of gravity.

The Moon's gravity is stronger on the side of Earth that happens to face the Moon than on Earth's other side. This difference in gravity is the dominant cause of the oceanic tides. Comet Shoemaker-Levy 9 broke apart because of the difference in Jupiter's gravity from one side of the comet to the other.

An object is more likely to be ripped apart if it comes very close to a strong source of gravity. If you do the algebra correctly, the equations of gravity show that Jupiter's tidal force on comet Shoemaker-Levy 9 was over six hundred times larger than the Moon's tidal force on Earth. The comet never had a chance.

Dear Merlin,

Does mass (or distance from a parent body, as in satellites of planets) govern rotational speed?

RICHARD E. HANSEN

MURPHYS, CALIFORNIA

Yes. Tidal interaction among orbiting objects such as the Sun and planets and planets and moons serves to slow the rotation rate of both members continuously until their rotation periods equal their orbital periods. Objects that have low mass and are in close orbits are most readily affected.

For example, the Moon is in tidal lock with Earth, and most of Jupiter's moons are locked with Jupiter. Depending on the shape of the object and the elongation of its orbit, other tidal locks are possible. Mercury is in a two-thirds tidal lock with the Sun. Its rotation period is exactly two thirds of its orbital period, so that three Mercurian days drag on for two years.

Dear Merlin,
   What is absolute zero?
   ANN CRONIN
   OAKLEY, CALIFORNIA

*I*t just happens that Merlin chatted with the British physicist
Sir William Thomson one Tuesday afternoon during the Paris
Exposition of 1900. (Sir William was raised to the peerage
in 1893, thus becoming Lord Kelvin, baron of Largs.)
Lord Kelvin attended the expo to deliver a popular lecture
on the physics of heat. Merlin was there to have a good time.
Merlin asked Lord Kelvin some questions while enjoying a
glass of 1893 Mouton at a nearby café.

MERLIN: Lord Kelvin, you have done much work on the theory
   of heat. Would you please explain the concept of absolute
   zero?

LORD KELVIN: With pleasure, Merlin. In my earlier work I
   advanced the idea that there exists a coldest possible
   temperature. Scientists now refer to this temperature as
   "absolute zero." It is the temperature that an object would
   have if all its heat were removed. I first made this
   suggestion based on the experiments of the French chemist
   Jacques Charles.

MERLIN: To make an object colder, couldn't you just add more
   cold?

LORD KELVIN: Heat exists. Cold does not exist. This is why you
   can raise the temperature of an object to no limits by

simply adding heat. It logically follows that to cool an object you must remove its heat.

MERLIN: One may conclude that an object cannot get colder when no heat remains. And that this is the coldest possible temperature.

LORD KELVIN: Absolutely!

MERLIN: Ha-ha.

LORD KELVIN: At this temperature, the object has reached zero on the absolute temperature scale. There are no negative temperatures. Just for reference, a toasty room temperature is about 300 degrees on the absolute scale. For convenience, I designed the scale so that each degree represents the same change in temperature as a degree on the common centigrade scale of Anders Celsius.

MERLIN: Maybe the absolute-temperature scale will one day be named for you.

In 1967, at the Thirteenth General Conference on Weights and Measures, Lord Kelvin received the ultimate (possibly dubious) honor of having the physical temperature unit "degrees Kelvin," which was already in common use, renamed to the simple lower-case "kelvin." He now joins watt, volta, ampere, joule, and others in having his name reduced to lower case. The Kelvin absolute-temperature scale continues to be used by nearly all the world's scientists.

Incidentally, zero kelvin equals −459.69 degrees Fahrenheit and −273.16 degrees Celsius.

Dear Merlin,

The production of light uses up the source of energy. Does the production of gravity consume the source?

PHILIP A. ROBERTS, JR., P.C.

CHESTERFIELD, VIRGINIA

The particles that carry gravitational energy are called gravitons. Merlin's good friend Albert Einstein first predicted their existence in 1916 as a consequence of an accelerating mass in the framework of his general theory of relativity. Gravitons as well as photons travel at exactly the speed of light.

The best examples of graviton production may be found among binary pulsars. The rapidly revolving variety are some of the best timekeepers in the galaxy. Based on their speed of revolution, one can compute the expected rate of energy loss due to gravitons. This is true in spite of the fact that the elusive graviton has never been detected in experiments.

The loss of energy is not revealed through the consumption of the star but through a drop in the energy of revolution. Indeed, the pulsar energies are dropping at exactly the expected rate—an important confirmation of general relativity.

IX

TIME
and
SPACE

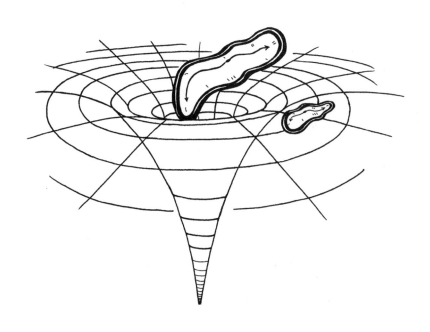

Dear Merlin,

What is the exact moment of sunrise or moonrise as it is given in the newspapers? Is it when the top, the middle, or the lower edge of the object emerges from the horizon?

GREGORY MILLER
SAN DIEGO, CALIFORNIA

By convention, and for no other reason, sunrise is the instant when the first tip of the Sun peeks above the horizon, and sunset is when the last tip of the Sun disappears behind the horizon. Moonrise and moonset are similarly defined.

The bending of light from the vacuum of interplanetary space to Earth's atmosphere allows you to see an image of the Sun's disk several minutes before it actually rises and several minutes after it actually sets.

Curiously, these combined facts of convention and atmospheric optics allow the published sunrise and sunset times to alter the correspondence (by several days) between the two calendar equinoxes and the two days of the year that have exactly twelve hours of light and twelve hours of darkness.

Dear Merlin,

Why will eggs stand straight up (balance) at a certain time during the vernal equinox in the Northern Hemisphere? Can this occur during the autumnal equinox? Also, are the conditions opposite this in the Southern Hemisphere?

JOHN BERKLICH, JR.
HIBBING, MINNESOTA

*E*ggs are no more likely to balance during the equinoxes than at any other time of year. Misconceptions abound because nobody thinks about balancing eggs on other days of the year. Those who succeed on March 21 promptly proclaim, "Eggs balance on end during the equinox" and thereby spread hard-boiled rumors.

Dear Merlin,

I heard somewhere that summer is longer than winter. Is this true? If it is true, then how can it be true? I thought the year was symmetric.

SONJA BENSON
BURDETT, NEW YORK

In spring and summer, Earth is farther from the Sun and travels more slowly in its orbit than in autumn and winter.

Alert calendar watchers (or people with nothing better to do) have always known that there are seven more days between the first day of spring and the first day of autumn than there are between the first day of autumn and the first day of spring. If you live in the Northern Hemisphere, this indeed means you have a longer summer.

You can credit the extra days to laws of planetary motion that were first realized in the early seventeenth century by one of Merlin's good friends, the German astronomer Johannes Kepler.

Dear Merlin,

Is the summer solstice the point where Earth is farthest from the Sun? If so, why should the two be coincident?

JOEL GRINBERG

SAN JOSE, CALIFORNIA

There are countless coincident cosmic happenings that are entirely unrelated. The summer solstice and Earth's aphelion (when Earth is farthest from the Sun) are among them.

But all this depends on what you mean by "coincident." The summer solstice occurs in the third week of June, while Earth's aphelion occurs two weeks later, in July.

Dear Merlin,

Is the amount of daylight per year the same at every point on Earth's surface? In other words, do the long summer days in Fairbanks, Alaska, balance out the long winter nights, giving the residents as much daylight per year as the people in Miami, Florida?

STEPHEN PENRICE
MESA, ARIZONA

The interval of sunlight during summer days in Fairbanks (located just south of the Arctic Circle) is indeed balanced by the interval of dark during winter nights in Fairbanks. But one needn't be in Fairbanks for this to be true. From the equator to the poles, Earth-dwellers average one-half year of sunlight and one-half year of dark.

This balance, however, is tipped slightly in favor of sunlight—by about one day—because of the refracted image of the Sun that projects over the horizon. This optical effect serves to add more sunlight to the days of the year than would be logged if Earth had no atmosphere.

Dear Merlin,

My question is about the precession of the equinoxes. I know that in 3000 B.C. the North Pole of Earth pointed to the star Alpha Draconis. Now it points to Polaris, and in A.D. 14,000 it will point to Vega. Earth is at a 23½ degree angle with the Sun. Now, when the precession changes on its 26,000-year circle, does it change the angle of Earth to the Sun? Maybe even to zero degrees?

EDWARD FOX

SALINAS, CALIFORNIA

There is always a lot of confusion about the behavior of Earth's axis.

All planets in the solar system have a tipped axis. The tip is the angle between a planet's axis and the vertical. This vertical (the solar system's "up") is defined by convention to be the direction that is at right angles to Earth's orbit, just as the plane of the solar system is defined by Earth's orbit itself.

The wobbly Earth sweeps 26,000-year circles around the vertical at an angle that ranges from 22 degrees to 24½ degrees. In the meantime, Earth orbits the Sun with an axis that for now points near Polaris. Over Earth's twelve-month orbit, the tilted axis alternatively finds itself tipped slightly toward the Sun (in the Northern Hemisphere summer), sideways to the Sun (during autumn and spring), and slightly away from the Sun (during Northern Hemisphere winter).

Unlike the planet Uranus, whose axis is tipped 98 degrees, Earth's axis will never point straight at the Sun.

Dear Merlin,

When does the twenty-first century officially begin? Is it January 1 of the year 2000, or is it January 1 of the year 2001?

Margaret M. Sipple
Pasadena, California

Calendar purists will tell you that the twenty-first century begins on January 1 in the year 2001. In the current Christian-based Gregorian calendar, the year 1 B.C. was followed by the year A.D. 1—there was no year zero. Century reckoning is therefore shifted by one year.

Before you get excited about this, please note a few things. (1) The "zero point" of any reckoning of time is arbitrary. (2) No day before October 15, 1582, actually occurred on the Gregorian calendar, for that was when the calendar was implemented officially. And (3), the Gregorian calendar doesn't even start in the "right" place. Bible scholars generally place the birth of Christ several years *before* 1 B.C.

When we consider the arbitrary numerics of it all, it shouldn't really matter which January 1 you choose to inaugurate the twenty-first century. Why not celebrate it twice?

Dear Merlin,

Knowing that Earth's year is 365 days long, an ancient astronomer may have divided it by 4, the number of Earth seasons. This would have given 91.25 days per season. By rounding this off to the nearest whole number you get 91, the product of 7 times 13. Could this possibly be the astronomical basis for the seven-day week?

MICHAEL KLAUS
CARROLL, IOWA

*N*o celestial phenomenon varies on a period of exactly seven days. Some West African tribes had a four-day week, while the Incas had a ten-day week. The seven-day week of the current Gregorian calendar can be traced to the combined influence of primarily two sources.

First, the early pagan civilizations of Mesopotamia believed that the heavenly bodies were gods and that the most powerful gods were the seven visible "planets" that moved among the fixed ones. They ranked these supergods by assuming that the more slowly they moved in the sky, the more ancient and powerful they were—thus ordering Saturn, Jupiter, Mars, Sun, Venus, Mercury, and the Moon. The Mesopotamians, and later the Romans, assigned seven-hour intervals of every day to each planet in succession. The planet (god) that ruled the first hour of each day was designated the ruler of the entire day, with the day named in the god's honor. With some pen and paper (or abacus), you can rederive their seven-day planet sequence: Saturn-day, Sun-day, Moon-day, Mars-day, Mercury-day, Jupiter-day, and Venus-day. Substitute "Sabbath" for Saturday and "Lord's day" for Sunday, and in Latin you get the basic forms for the French, Italian, Spanish, and Portuguese names for the days of the week: Sabbata, Domenica, Luna, Martis, Mercurius, Jovis, Veneris. In English, substitute the Saxon gods Woden, Thor, and Frigga to get Wednesday, Thursday, and Friday.

Another source of the seven-day week is traceable to the Israelites and the Book of Genesis in the Old Testament: "And on the seventh day God ended his work which he had made."

Dear Merlin,

How accurate is the leap-day correction to the calendar every four years?

CAL LOBEL
NEW YORK CITY

*N*ot very.

Earth takes slightly less than 365 days and 6 hours to complete its seasonal orbit. In a world without leap days, the vernal equinox (first day of spring) would slide backward through the days of March at a rate of about a day every four years. There would be great unrest in the fashion industry when the seasons lost their correspondence to the calendar months.

The solution: every four years donate twenty-four hours to February, the neediest month (which is the scheme of the Julian calendar)—but this overcorrects the problem by what accumulates to be slightly less than a day every one hundred years.

The solution: every 100 years (e.g., 1700, 1800, 1900), omit the leap day that would have otherwise appeared—but this scheme undercorrects the problem by what accumulates to be slightly less than a day every 400 years.

The solution: every 400 years (e.g., 1600, 2000, 2400), reinsert the leap day that would have otherwise been omitted. Behold the Gregorian calendar, introduced to the Western world by Pope Gregory XIII and his Calendar Commission. The new calendar was jump-started in 1582, when the day after October 4 was decreed to be October 15.

But the Gregorian calendar itself overcorrects the problem.

The solution: subsequent calendar reform now mandates that every 4,000 years (e.g., 4000, 8000, 12,000), the leap day that would have been added by all previous rules should be omitted—but this undercorrects the problem, which obviously can be solved when you reintroduce a leap day in every century year that leaves a remainder of 200 or 600 when divided by 900.

You may now rest peacefully, knowing that the first day of spring will barely budge for the next forty-four millennia.

Dear Merlin,

What are the lengths of a universal day and a galactic day in Earth-day terms?

Cary D. Nolan

San Jacinto, California

The universe has never been measured to rotate, so one cannot meaningfully translate a universe day into an Earth day. But a galaxy day, which one might interpret as a full revolution at the location of the solar system, drags on for about 80 billion Earth days.

Dear Merlin,

How long does it take for light to reach Earth coming from the Sun?

MICHAEL OYE

OXNARD, CALIFORNIA

*I*t takes the Sun's light and its gravity about 495 seconds in January and about 505 seconds in July to reach Earth. These months are when Earth, in its elliptical orbit, is closest to and farthest from the Sun.

Yes, it is true. If somebody yanked the Sun from the solar system, then you and Earth's orbit would remain clueless for over eight minutes.

Dear Merlin,

My question relates to time. I am told that time began at the beginning of the universe, but if time had a beginning, then what was "time" before it began? And if time were to end, wouldn't there still be "time" after it ended?

PAUL W. QUANDT

AUBURN, CALIFORNIA

*W*here there are no repeating phenomena, the measurement of time, and time as we know it, cannot be defined.

The interval that separates two events is commonly measured by something that repeats—the vibrating crystals of a modern digital watch, the pendulum and gears of a grandfather clock, the rising and setting Sun, heartbeats, etc. We presume that outside the universe (this includes before the beginning and after the end), there are no events and, of course, no measuring devices. In such a place, time would have no meaning. In practical terms, time would be irrelevant.

But do not let this upset you. There are many systems where time is mostly irrelevant, and you needn't leave the universe to find them. All those who use mathematics to describe the physical universe will use time in their equations only when something changes in their system. Otherwise, the system is in "steady state," and time is handily excised from the computations. For example, the structural engineers who design today's suspension bridges have few occasions to use time in their equations. A bridge is the graceful conquest of balanced forces. Were it not for corrosion, continental drift, and the Sun becoming a red giant, a bridge would last forever—in timeless, blissful, and steady state.

Dear Merlin,

If the light generated from an event that occurred more than two hundred years ago were to travel one hundred light-years into space, bounce off some space object, and return one hundred light-years back to Earth, would it be possible to "watch"—say, the signing of the Declaration of Independence?

SUSAN JONES

MIDLOTHIAN, VIRGINIA

*I*f the Declaration of Independence were signed in an open field, and if there were no obscuring clouds in the sky, and if you floated special image reflectors more than a hundred light-years away in space, and if you constructed a jumbo telescope (larger than the diameter of Earth), you could easily recover some of the original light and watch this historic event.

By the way, Spica (Alpha Virgo), which is the brightest star in the constellation Virgo, is at just about the right distance to watch two-hundred-year-old Earth events in real time. And Merlin's relatives, two million light-years away in the Andromeda galaxy, are now watching the end of Earth's Pliocene epoch.

Dear Merlin,

I am confused. When we look to the most distant objects in space, we see light coming to us from a time when the universe was young. How did Earth get ahead of the light so we can look back and see it coming to us?

ED SCHNUTENHAUS
SAN JOSE, CALIFORNIA

*W*hen you look far away, you do indeed look far back in time. You see a sample of the universe not as it is but as it was—the light emitted long ago, in spite of its blazing speed, has only just now reached our telescopes. But not all of the early universe is available for you to see. Part of it evolved to become the Milky Way and the entire Local Group of galaxies.

During the era of quasars, for example, only some of them were far enough away from the youthful Milky Way for their light to be reaching us now. Those quasars that were nearby must have eventually stopped being quasars. How do we know? Because there are no nearby quasars today. Merlin is not making this up. There is mounting evidence that supermassive black holes discovered in the centers of nearby galaxies may be the dead engines of former quasars.

Dear Merlin,
  What is space itself?
  JAMES LAURITS
  APTOS, CALIFORNIA

[Space] is a new ocean, and I believe the United States must sail upon it.

JOHN FITZGERALD KENNEDY

Nature abhors a vacuum.

A BATHROOM WALL

Distance extending without limit . . . within which all material things are contained.

WEBSTER'S NEW ENCYCLOPEDIC DICTIONARY
OF THE ENGLISH LANGUAGE

Matter tells space how to curve. Space tells matter how to move.

JOHN ARCHIBALD WHEELER

Every cubic inch of space is a miracle.

WALT WHITMAN

Space [is] the final frontier.

CAPTAIN JAMES T. KIRK, USS ENTERPRISE

Space is the next frontier.

MERLIN

Dear Merlin,
    What are the first words ever spoken from space?
    SHOHN TROJACEK
    ENNIS, TEXAS

The first words from space are unquestionably "bleep, bleep," as broadcast from the Soviet satellite *Sputnik 1*, launched on October 4, 1957.

If you do not believe that hardware should hold the first-words-from-space record, then the first words are "woof, woof," as barked by the Russian dog Laika aboard the second Soviet satellite, *Sputnik 2*, launched on November 3, 1957.

If for some reason you feel that the record should be held only by a *Homo sapiens*, then the first words are "Поехали, поехали!" ("We ride, we ride!"), proclaimed by the Russian cosmonaut Yuri Gagarin as the engines of *Vostok 1* thrust him into orbit on April 12, 1961.

Dear Merlin,

I was wondering, since helium balloons float, would helium balloons float in outer space? I would also like to know if they would float on the Moon.

DEVERI HANSEN

LIBBY, MONTANA

*H*elium is simply lighter than air. On the airless Moon, your otherwise buoyant helium balloon would fall like a brick to the surface.

And yes, a helium balloon will float in space, but so would a balloon filled with lead pellets. As you drift through outer space, not only can nobody hear you scream, nobody can measure your weight either.

Dear Merlin,

Let's make believe that I am located directly between Earth and your home planet, Omniscia. Spaceships routinely pass by at the speed of light on their journey between the planets. To me, the occupants of the spaceships age more slowly than I do, and the lateral dimensions of the spaceships appear to shrink. I am most curious to know what an observer in a spaceship traveling from Earth would observe about the spaceship traveling toward Earth in the above scenario.

RON GIFFIN

CINCINNATI, OHIO

There is no known law of physics that permits a spaceship to travel at the speed of light. For your example, let's slow them down a bit, to 99 percent of the speed of light.

Naive thinking would say that one spaceship would clock the other passing at 198 percent of the speed of light, since all motion is relative. For speeds near the speed of light, however, Einstein's special theory of relativity demands the use of a more complicated equation to compute the veritable circus of relativistic effects.

Merlin's abacus shows that each spaceship will (1) see the other spaceship going past at 99.995 percent of the speed of light, (2) notice that time on the other spaceship progresses at one one-hundredth of the usual rate, (3) measure the mass of the other spaceship to be one hundred times its usual mass, and (4) notice the other spaceship to be one one-hundredth of its normal length.

Dear Merlin,

Einstein says that space is curved and that if you were to travel forever, you would come back to where you started. Okay, even if that were true, wouldn't there be a vacuum of space beyond that curve, and wouldn't space have no ending and go on forever even though the human mind could not conceive it?

PAUL W. QUANDT

AUBURN, CALIFORNIA

A t the root of your question is a deeper query: Does "nothing" enclose the universe?

The best vacuum you will find anywhere, according to four out of five vacuum retailers (and five out of five astronomers), is the void of intergalactic space. But we can then ask, "Is intergalactic space 'nothing'?" No, it still contains space.

If you feel obliged to call intergalactic space "nothing," then you must invent a word to refer to the region outside the universe. In this location, where we presume there to be no space, there can be no nothing. Let's call it—we are left with no choice—"nothing-nothing."

X

Our Universe

Dear Merlin,
   What is the universe?
   Rose Putnam
   Shelley, North Carolina

Since we know of only one universe, we cannot make comparative statements, any more than an infant in a mother's womb can compare itself to other unborn infants.

We must be content with the almost metaphysical statement "The universe is the contents of all the space and time in which we can exchange information." By this reasoning, for example, when you take that one-way trip into a black hole, you have effectively left the universe.

Dear Merlin,

What I'd like to know is, what is the exact size and shape of the universe, and does it have an outside?

ALVARO B. GALLI
BROOKLYN, NEW YORK

You are not alone. Merlin would also like to know the answers to those questions.

Dear Merlin,

   Are there really more stars in our galaxy or universe than grains of sand on Earth?

BILL DELLINGES

NEWARK, CALIFORNIA

$\mathcal{S}$ome sandy facts:

1. Depending on how much you play on a beach, you might dig more grains of sand out of your bathing suit than there are stars in some galaxies.
2. But let it be known, there are more stars in the universe than grains of sand on an average beach.
3. Yet in the end, grains of sand win. Add up the grains of sand on all Earth's beaches, in all Earth's deserts, and below Earth's oceans and you have easily outnumbered all the stars in the observable universe.

Dear Merlin,

Astronomers say that there are 100 billion stars in our galaxy and possibly 100 billion galaxies. This would total the incredible number 10 billion trillion stars. Is there a scientific method to determine such a number? If so, what is it?

HAROLD R. HODGSON

HANOVER, PENNSYLVANIA

Big astronomical quantities are easier to calculate than you might think. A method is used that is not much different from what you might use at one of those jelly-bean contests where you are asked to count the jelly beans contained in a jumbo jar. People who are in a hurry will (1) count the beans in a minivolume within the jar, (2) guess how many minivolumes will fill the whole jar, and (3) multiply the two numbers to get an estimate of the total number of beans.

We know approximately how many stars there are in a minivolume around the solar system, and we know approximately how many of these minivolumes would fill the galaxy. When you multiply these two numbers, you get something in the vicinity of 100 billion stars.

We know approximately how many galaxies there are in a minivolume around the Milky Way, and we know approximately how many of these minivolumes would fill the universe. When you multiply these two numbers, you get as many as 100 billion galaxies.

At the risk of becoming starstruck, we now multiply 100 billion stars per galaxy by 100 billion galaxies to get 10 sextillion (10 billion trillion) stars.

And just in case you were wondering, 10 sextillion jelly beans would fill a jar that had a volume of about 2.4 million cubic miles.

Dear Merlin,

I once read: "That there's no night and day on Earth proves the universe has not existed for an eternity. So contend the astrophysicists." Why?

GORDON RIDLEY

SAN FRANCISCO, CALIFORNIA

If your question is "Why do the astrophysicists contend that there is no night and day on Earth," the answer is "They don't." If your question is "Why did my newspaper report that astrophysicists contend there is no night and day on Earth," the answer is "You should read a different newspaper."

The article was probably trying to describe Olbers's paradox, named for Heinrich Wilhelm Matthäus Olbers, an acquaintance of Merlin's who, in 1826, correctly reasoned that if the universe were infinite and if stars were spread evenly throughout all of space, every part of the daytime and nighttime sky would be ablaze with light.

Why? Distant stars are dimmer than nearby stars, but the number of distant stars is higher than the number of nearby stars. These two effects balance each other and leave constant the total intensity of light that reaches Earth from any chosen distance. If you had an infinite universe, there would be no shortage of light available to reach Earth. But the last time Merlin checked, the sky does get dark at night, so at least one of Olbers's assumptions must be false.

While whole galaxies of stars are found almost everywhere we look in the universe, the universe is neither infinitely large nor infinitely old—and it is expanding. Each of the three effects, taken alone, resolves Olbers's paradox. Taken together, you can sleep peacefully, knowing that it will be forever dark at night.

Dear Merlin,

An argument against an infinite universe is that the infinity of light from the infinity of stars would result in a bright uniform glow throughout the sky, which is contrary to observations. I understand this is known as Olbers's paradox. Is it possible for the universe to be filled with light-absorbing black holes in such a way that beyond a certain distance no light would reach us?

RON GIFFIN

CINCINNATI, OHIO

A space-filling presence of black holes would be detected easily by the gravitational "lensing" of background light from distant galaxies and quasars. Separate light paths that would otherwise miss us are bent along the curved space in the vicinity of a black hole. What we see is a split image of a single object. Dozens of lensed galaxies and quasars are known to exist. But a space-filling presence of black holes would "lens" them by the thousands, which is not observed.

Dear Merlin,

How many atoms are in the known universe, and what sort of bona fide number could be attached to the total?

JACK LARNED

SAN ANTONIO, TEXAS

*E*arly in this century, the famous English astronomer Sir Arthur Eddington, another acquaintance of Merlin's, once claimed he knew exactly how many atoms there are in the universe. But unless you fly around the cosmos to count them, the exact number of atoms is not readily knowable. For good reason, Sir Arthur is better known for his theoretical investigations of stellar structure than for his numerology.

An estimate of the total number of atoms in the universe is remarkably easy to compute. The mass of the typical atom (over 90 percent of the atoms in the universe are hydrogen atoms) is well known, as is the mass of the average star and the number of stars in an average galaxy. When combined with the latest estimates for the number of galaxies in the universe, Merlin's abacus shows about 10,000,000,000,000,000, 000,000,000,000,000,000,000,000,000,000,000, 000,000,000,000,000,000 total atoms—give or take a few.

This number has no official name (although Merlin votes for "totillion"), but it can be written more compactly as ten raised to the seventy-ninth power: 10^{79}.

Dear Merlin,

I am six feet tall. Comparatively speaking, which is the farthest away from my eye, the atomic particles in my little toe or the most distant galaxies?

JACK LARNED

SAN ANTONIO, TEXAS

Since you are only six feet tall, there is no doubt that the distant galaxies are farther from your eyeballs than your pinky toe.

Another way to address your inquiry, however, is to imagine you suddenly became ten times taller, then ten times taller again, and so forth. After about twenty-five of these stretching exercises, your head would brush up against the most distant galaxies in the visible universe. Returning to Earth, imagine you shrink in successive steps so that each step takes you one tenth of the distance to the atoms in your toe. It would take only fifteen shrinks to enter the nuclei of your toe atoms.

Yes, the entire known universe spans about forty of these jumps.

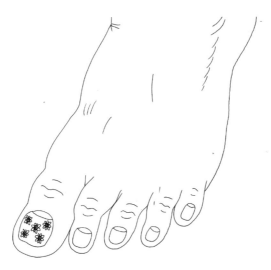

Dear Merlin,
What is Einstein's "cosmological constant"?
MARY LEE COLEMAN
PARSIPPANY, NEW JERSEY

*E*instein's general theory of relativity, published in 1916, contains equations that favored a universe that was either expanding or collapsing. At the time, the concept of an expanding or collapsing universe was farfetched. Realizing this potential folly, Einstein introduced a mathematically legitimate antigravity "pressure" term to his equations—the cosmological constant—which allowed the universe to remain a constant size.

Thirteen years later, when Edwin Hubble showed that the universe was indeed expanding, Einstein quickly removed the cosmological constant and declared it to be his life's greatest blunder.

Little could he have predicted that the cosmological constant would be resurrected as a tool for the modern cosmologist to reconcile inconsistencies between the age of the universe when estimated from its stars and from its current expansion rate.

Dear Merlin,

What is the "Hubble constant"? Why is it constant? Since I see it often in astronomy articles, I know it is important. But can you tell me why it is important?

JAMES O. ROBERSON
WILLIAMSTON, NORTH CAROLINA

*M*erlin chatted with the American astronomer Edwin Powell Hubble shortly after his research on the expansion of the universe was published in the *Proceedings of the National Academy of Sciences.* Merlin and Edwin sipped tea in a library near Edwin's favorite telescope on Mount Wilson, outside Pasadena, California.

MERLIN: Ed, what should curious people know about your recent work?

HUBBLE: My colleague Vesto Slipher at Lowell Observatory in Arizona has made available the spectra of forty-one galaxies. I used the excellent hundred-inch Mount Wilson reflecting telescope here to measure the distances to eighteen of them.

MERLIN: The year is 1929. Isn't this Mount Wilson telescope the largest in the world?

HUBBLE: Indeed it is. The galaxy spectra contain shifted features that permit us to determine a velocity for each of them. Vesto showed that nearly all the galaxies measured were moving away from our Milky Way galaxy, as though the universe were expanding.

MERLIN: How about the Andromeda galaxy?

HUBBLE: That's your home, isn't it? Well, Andromeda and just a few other very close galaxies move toward the Milky Way.

MERLIN: For the rest of the galaxies, what did you find?

HUBBLE: I obtained distances for most of them, and it seems that the farther galaxies recede from us faster than the near galaxies, at a constant rate of about 100 miles per second for every million light-years distant.

MERLIN: This means that galaxies at one million light-years recede at 100 miles per second, and galaxies at two million light-years recede at 200 miles per second, and so forth?

HUBBLE: Yes, Merlin. And remarkably, some new data from my coworker Milton Humanson here at Mount Wilson show that the rate appears to be constant out to great distances, although I do not know if it has been constant for all of time.

MERLIN: Perhaps one day it will be called the "Hubble constant" in your honor.

HUBBLE: Perhaps. More tea?

With newer and bigger telescopes and with more and better data, the Hubble constant has been revised to be considerably smaller than first estimated by Hubble himself. The latest, best value is about 15 miles per second per million light-years. Note that in the big-bang scenario, the expansion rate of the universe (and therefore the Hubble constant) is constant across space but not constant over time.

Dear Merlin,

I understand that in an expanding universe, an important factor in determining the speed of a receding galaxy is the redshift observed in the galaxy's spectrum. The galaxies that recede with the greatest speed are also the most distant.

Suppose the redshift were caused in part by photons losing energy through collisions with atomic and subatomic particles on their way to Earth. Is a redshift possible due to a quantum loss of energy in these collisions, and if so, would this then require a reevaluation of the apparent size of the universe?

CHARLES J. MUELLER

CHICAGO, ILLINOIS

You have mentioned what is commonly called the tired-light model. Galaxy light can be reddened in the way you describe, but this is not equivalent to a Doppler shift, where well-defined spectral features are bodily shifted to another place in the spectrum.

In the tired-light model, every photon (of all wavelengths) must lose exactly the same fraction of its energy over the same travel distance. The physics of collisions and scattering indicates that a photon could get tired, but it would also get dizzy. The energy shift would be irregular and thus blur what are otherwise observed to be sharp spectral features.

Many noble attempts have been made to salvage the tired-light model, but galaxy Doppler shifts are still most simply and best explained by an expanding universe.

Dear Merlin,

Is it true that everything from quarks to stars in the whole blinking universe rotates?

T OM D IX
O CEANSIDE, C ALIFORNIA

Subatomic particles have a property called spin, which is analogous to rotation. Everything else in the universe can be said to turn on an axis, although there exists a big range in rotation speeds. Systems in which rotation is often unimportant to their properties or structure include gas clouds, star clusters, and many dwarf galaxies. Systems in which rotation is important include pulsars and highly flattened objects such as spiral galaxies and pizza dough.

Dear Merlin,

Since the solar system and the galaxy have distinct axes, I am curious to know whether the universe has an axis of rotation.

ROBERT SIEGEL, M.D., PH.D.

PALO ALTO, CALIFORNIA

There is no evidence that the space of the universe has a center, an axis of symmetry, or an axis of rotation.

The turn-of-the-century Austrian philosopher Ernst Mach (fast travelers know that the speed of sound is named for him) postulated that mass, or more generally inertia, could be a consequence of the centrifugal forces within a rotating universe. As intriguing as this suggestion sounds, and as influential as Mach's ideas have been, detailed experiments have shown that Albert Einstein's description of gravity and inertia provides a deeper and more accurate understanding of the universe.

Dear Merlin,

If the universe were spinning, how could we tell?

R*OBERT* H*ANAN*, M.D.

S*AIPAN*, M*ARIANA* I*SLANDS*

If the entire universe rotated, it would go unnoticed only if your Milky Way galaxy were at the center of all rotation. Redshift surveys of galaxies detect motion toward or away from the observer; the component of a galaxy's motion that is sideways to your view is unmeasurable by all existing observational methods.

If the Milky Way were positioned anywhere other than at the center, then the "smoking gun" of a rotating universe would be large sections of the universe approaching you, other sections receding from you, and still other sections neither approaching nor receding. Galaxy redshift surveys do not support this contention.

Dear Merlin,

Is the 3-degree cosmic background radiation of the big bang the same amount in every direction from Earth? And if so, does that mean we're at the center of the universe?

WILLIAM FRAY

LONG BEACH, MISSISSIPPI

When you adjust for the motion of Earth and the solar system within the Milky Way galaxy, the background radiation from the big bang (measured to be 2.726 degrees Kelvin) is remarkably constant in every direction. This would be true when viewed anywhere else in the universe. A unique center cannot be defined.

Dear Merlin,

Where is the theoretical center of the universe in the big-bang theory?
TASKER N. RODMAN, M.D.
LEACHVILLE, ARKANSAS

*W*ith the advent of relativity theory in the early twentieth century, Albert Einstein and Hermann Minkowski showed mathematically that time is as much a dimension as space. To ask "where?" is also to ask "when?" To help digest the proper answer to your question, we should first think in three dimensions—two space dimensions and one time dimension.

Imagine galaxies etched on the two-dimensional surface of a balloon. As the balloon expands, Merlin can ask you, "Where is the center of the balloon?" You will probably say in the middle—inside the balloon. But the balloon's surface, where we displayed our etchings, was at the center only when you first started to inflate the balloon. The center of the balloon does not exist on its surface, except back in time at the balloon's origin.

By comparison, in our four-dimensional universe—three space dimensions and one time dimension—you ask, "Where is the center?" Merlin can legitimately reply, "It happened everywhere in space at $T = 0$, the beginning of time and the beginning of the universe."

Dear Merlin,

How do the very small thermal variations in the cosmic background radiation account for the complex positioning of galaxies in our universe?

THOMAS LOUGH

OCEANSIDE, CALIFORNIA

Those tiny variations in the cosmic background contain the imprint of the nascent mass structure of the universe. In the jargon of the trade, the thermal variations betray the last point of scatter for the radiation before it decoupled from matter. Thereafter, gravity became more important, and the mass structures became more pronounced until they formed the large assemblies of galaxies we see today.

If galaxies were observed to be spread evenly throughout the universe, one would not expect significant variations in the cosmic background.

Dear Merlin,

Regarding the big bang, please explain how Earth alone, never mind all the matter in the universe, could be at a point infinitesimally small.

Tom H. Dix

Oceanside, California

*I*f there were no empty space within and among atoms, then the entire Earth would fit neatly into a sphere about a mile across. Ordinary physics can describe this scene without much trouble. If you allow matter to be converted to energy, you can pack as many photons in as tiny a volume as you please.

But one can reach a conceptual level in which ordinary physics leads to astonishing conclusions. Within the event horizon of a black hole, for example, equations that are based on well-understood principles of physics show that all matter collapses to an infinitesimal point. Similar equations and physical principles tell us that the universe was infinitesimally small at the moment of the big bang over ten billion years ago.

Either the universe really was infinitesimally small, in which case we can all walk around with boggled minds, or an undiscovered physics exists that prevents it. Large particle accelerators are designed in part to simulate the energies of the early universe and to explore the new physics that may lurk at the limits of common sense. Stay tuned, mind-boggled or otherwise.

Dear Merlin,

If the universe is "closed" and eventually recollapses owing to its own gravity, what will happen to the radiation energy that now fills space?

GEORGE LOCKWOOD

DAYTON, OHIO

*A*s the universe collapses, the "energy density" of space will systematically rise. The nighttime sky will eventually be ablaze with light, because in a total collapse, the 3-degree cosmic background radiation will become the 4-degree background, the 5-degree background, and so forth through the trillions of degrees.

Some checkpoints to look for: At a 3,000-degree background, the sky will glow red hot. At a 6,000-degree background, the sky will glow nearly white hot and be indistinguishable from the Sun's disk. And at a 30,000-degree background, the sky will glow a pretty blue once again (but for different reasons), but by then Earth will have been vaporized.

Dear Merlin,

I often read the physicists' attempts to describe the events immediately following the big bang: events during the first three minutes, the first second, or even the first tiny fraction of a second. I am unsure, however, what it means to be talking about these time periods, because as general relativity tells us, time slows down in the presence of gravity. Presumably the near-infinite gravity presented by the near-infinite density of the big bang forced time to slow to near zero. So what was a "second" in terms of elapsed time then? A trillion years as we know it?

DAVIS COHEN

AUSTIN, TEXAS

In a strong gravitational field, time will indeed be observed to slow down. This time-slowing, however, is not perceived at all by those within the strong gravity field. The phenomenon is relative, which means, in the case of your question, it can be measured only by an observer somehow placed outside the universe.

Without an external observer, the question of how long the beginning of the universe took is relegated to the timekeeping devices of the participants.

Dear Merlin,
   How will the universe end?
   S. KENT WALLACE
   GRAND RAPIDS, MICHIGAN

$\mathcal{M}$erlin asked that question of the British author T.S. Eliot in 1925. He replied with poetic clairvoyance:

> *This is the way the way the world ends*
> *Not with a bang but a whimper.*

Latest evidence suggests that there is not enough mass and gravity to halt or reverse the expansion of the universe. In this scenario, star formation ultimately consumes nearly all interstellar gas. The remaining stars exhaust all their fuel as the universe expands forever and cools to a dark, frigid, thermodynamic death. After about 10^{32} years (10 sextillion times the current age of the universe), even the proton, the very fabric of matter itself, will decay.

Such a fate may sound unpleasant, but if the universe recollapses, that will be equally horrific for its inhabitants. The scene will simply make better headlines as the universe shrinks back into the cosmic maelstrom that was the fireball from which it was born.

XI

ONE STEP
BEYOND

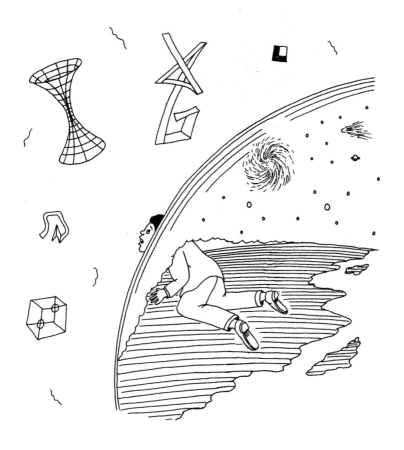

Dear Merlin,

A wandering philosopher has informed me that our entire observable universe is contained in drops of sweat that are rolling down a giant's back. What to us is millions of years is but a second to the giant. Other galaxies are other drops of sweat. A black hole is when a drop falls off the giant's back. I thought that you would be interested in this view.

JAMES J. CONTRADA
MALVERNE, NEW YORK

Sorry, but Merlin is not interested in that view.

It seems that your wandering philosopher has wandered too far. Such stories are the entertaining substance of mythology, not science.

Dear Merlin,
What is the meaning of life?
GREGORY LEWIS
NEW YORK CITY

*L*ife is that property of plants and animals which makes it possible for them to take in food, get energy from it . . . and reproduce their kind.
WEBSTER'S NEW ENCYCLOPEDIC
DICTIONARY OF THE ENGLISH LANGUAGE

By meditation, in the self see
Some the self by the self;
Others by discipline of reason,
And others by discipline of action.
BHAGAVAD GITA,
CHAPTER XIII, VERSE 24

Keep thy heart with all diligence; for out of it are the issues of life.
PROVERBS 4:23

The actuality of thought is life.
ARISTOTLE

There are but three events in a man's life: birth, life, and death. He is not conscious of being born, he dies in pain, and he forgets to live.
JEAN DE LA BRUYÈRE

Be ashamed to die until you have won some victory for humanity.

HORACE MANN

You never know what life means till you die;
Even throughout life, 'tis death that makes life live.

ROBERT BROWNING

Along a parabola life like a rocket flies,
Mainly in darkness, now and then on a rainbow.

ANDREI VOZNESENSKY

Life is a beach.

BATHROOM WALL IN CALIFORNIA

Life is that small part of the cosmic calendar where you take your glimpse of the universe.

MERLIN

Dear Merlin,

This question has nothing to do with astronomy, but since you are grounded in physics and philosophy I would like your opinion on the following matter. The scene is in a forest. A giant tree falls. No sentient being is within miles of this event. Question: Is there a noise when this tree falls? Some will say there is no noise unless there is an ear or receiving organ to perceive the noise and since there is none, there is no noise. What do you think?

ANDREW J. SMATKO, M.D.

SANTA MONICA, CALIFORNIA

$\mathcal{M}$erlin has three comments for you.

1. If no sentient being registered the event, then you did not register the event (assuming you are a sentient being). Since you did not register the event, you do not know at all whether the tree fell. Since you do not know if the tree really fell, you have no business asking the question.

2. In physics we define sound as the longitudinal vibration of atoms or molecules. Of course the tree made sound.

3. People who like to start arguments say that sound can be only what you hear and what your brain interprets as sound. Not only does this insult the canine community (since canines hear much higher pitches than humans), but it ignores the utility of nonsentient electronic detectors that are built to hear what we cannot hear and see what we cannot see.

Dear Merlin,

This is not astronomical, but could be, because of the mathematical grasp necessary to analyze and assign a probability to this true story.

In the express line of a supermarket, the man ahead of me bought the same product I had. That is to say, he had one item and I had an identical item. What are the odds against this happening? Make any reasonable assumptions necessary.

STEPHEN B. CHARNEY

PACIFIC PALISADES, CALIFORNIA

The odds are slim if the item was a tube of anchovy paste.

Otherwise, if the express lane is labeled "ten items or less," let us assume there is a 1-in-10 chance that the person ahead of you in the express lane is buying one item. If the store sells about 100,000 items, then the chances that the person has a single item that matches yours is $1/10 \times 1/100,000 = 1/1,000,000$, or one in a million. In practice, express lane buyers select from a relatively narrow assortment of items, such as milk, cigarettes, bread, and potato chips. If the "effective" store inventory were lowered to about 100 items, then the probability becomes $1/10 \times 1/100 = 1/1,000$, or one in a thousand. If you go to the express checkout lane once per week, then you can expect your event to happen about once every twenty years.

Dear Merlin,

I watched a film regarding the end of the world, scheduled for May 5, 2000. All the planets and the Sun are in line, and the South Pole is shifted, etc. Have you heard about this philosophy, which someone discovered in the 1940s?

C. WARREN CULLAR

AUSTIN, TEXAS

On May 5, 2000, the Sun, Moon, Mercury, Venus, Mars, Jupiter, and Saturn are within 30 degrees of each other on the sky. But before you jump to cataclysmic conclusions, you should know that such a configuration is not rare. The last time it happened was on November 15, 1982, and it has happened fifteen other times in the twentieth century. Did anybody make a film about the end of the world then? Probably not. But even if they had, clearly the world survived the moment.

Also note that whenever the "big seven" come together in the sky, the event is observationally forgettable. The Moon is unobservable as it nears its new phase, and the rest of the planet jamboree is mostly lost in the glare of the Sun.

The new millennium will come and go, and the world will still be here, whether or not people make money preaching otherwise to an unsuspecting public.

Dear Merlin,

Do you (and other scientists) allow for the theoretical possibility of the existence of God? Is there room for both a scientific and a pseudo-scientific explanation of the origin of the universe?

SHORTI THOMPSON

BEDFORD, VIRGINIA

To allow for the existence of God or to invoke pseudo-science in the origin of the universe is to allow a system of faith-based belief that plays no role in scientific deduction.

In the predominant religions of the world, the existence of God and related paranormal phenomena (such as miracles) are held to be true on faith. Faith is also a source of spiritual reflection and enlightenment. This is why religions are collectively called faiths. Indeed, religious people who seek scientific evidence to support their faith ought to reconsider how religious they really are.

The basis of modern science is the scientific method. When properly applied, the scientific method *requires* experiment to verify (or falsify) a belief. Claims about the universe that cannot ever be tested are of little use to scientific search and discovery. This simple fact holds as true for "God created the universe" as it does for "Invisible extraterrestrials live among us in a parallel universe."

Theories of the universe are as old as human consciousness. Only in this century have true scientific theories for the origin of the cosmos been formulated—primarily because of Albert Einstein and large telescopes. The big-bang theory is the latest and best tested among them.

Dear Merlin,

I have a pretty good idea of what pi is. It came from the ancient Greeks and it is the ratio of the circumference of a circle to its diameter. I also know that it is a never-ending decimal, 3.14 . . . How do we find each exact digit after the decimal of pi?

JASON DUGGAN

STAMFORD, TEXAS

$$pi = 3\tfrac{1}{8}$$

EGYPTIANS (c. 2000 B.C.)

$$pi = (\tfrac{16}{9})^2$$

BABYLONIANS (c. 2000 B.C.)

$$pi = 3$$

BIBLE (c. 550 B.C.)

$$3\tfrac{10}{71} < pi < 3\tfrac{1}{7}$$

ARCHIMEDES (287–212 B.C.)

$$pi = 2 \times (\tfrac{2}{1} \times \tfrac{2}{3} \times \tfrac{4}{3} \times \tfrac{4}{5} \times \tfrac{6}{5} \times \tfrac{6}{7} \times \tfrac{8}{7} \times \tfrac{8}{9} \times \ldots)$$

JOHN WALLIS (1616–1703)

$$pi = 3 + 3 \times \tfrac{(1/2)^3}{3} + 3 \times \tfrac{3}{4} \times \tfrac{(1/2)^5}{5} + 3 \times \tfrac{3}{4} \times \tfrac{5}{6} \times \tfrac{(1/2)^7}{7} + \ldots$$

ISAAC NEWTON (1642–1727)

$$pi = 4 \times (1 - \tfrac{1}{3} + \tfrac{1}{5} - \tfrac{1}{7} + \tfrac{1}{9} - \tfrac{1}{11} + \tfrac{1}{13} - \tfrac{1}{15} + \ldots)$$

GOTTFRIED LEIBNIZ (1646–1716)

$$\text{pi} = \sqrt{8 \times (\tfrac{1}{1^2} + \tfrac{1}{3^2} + \tfrac{1}{5^2} + \ldots)}$$
LEONHARD EULER (1707–1783)

pi = 3, exactly
A MISGUIDED BUT WELL-INTENTIONED
NORTH CAROLINA STATE LEGISLATURE (1932)

pi = 3.14159265359 . . .
PI BUTTON, HANDHELD CALCULATOR (LATE 20TH CENTURY)

Dear Merlin,

    Using the resources and technology of today, is it possible to send astronauts at 10 percent of the speed of light or greater?

    DARYL KOHLERSCHMIDT

    BAYFIELD, COLORADO

*M*ost decidedly no.

Ten percent of the speed of light is over 2,600 times faster than the maximum speed reached by the *Apollo* astronauts in the 1960s and 1970s en route to the Moon. Consider that this top speed is only about 800 times the speed of the original *Wright Flyer*, the Wright brothers' first plane flown in 1903 at Kitty Hawk, North Carolina. Incidentally, Merlin was good friends with Wilbur and Orville.

    Merlin does not predict the future, but speculation based on the physical principles of today can be instructive. The resources and technology of tomorrow may bring low-mass impulse rockets that are large enough to propel astronauts. One version uses ions: if you strip an atom of one or more of its electrons, you can use a magnetic field to accelerate the atom and eject it from the rear of the spacecraft. The spacecraft then recoils forward. Another version uses laser photons: a ship moves forward with the impulse it receives from emitted particles of light. A more ambitious idea is the hydrogen ram jet, which would scoop up unsuspecting hydrogen atoms in interstellar space to feed an on-board nuclear fusion reactor.

    Unlike conventional chemical fuel, these impulse drives can send an astronaut to near the speed of light, but the acceleration is very slow. You would not come near the speed of light for thousands of years. If you were to travel anyplace interesting, then by the time you arrived you would be long dead.

The centuries to come may also bring a new technological understanding of the fabric of space-time where the distance between you and your destination may be lessened by an induced curvature of the intervening space. This idea was exploited for the "warp drives" of the starships in the popular American television (and film) series Star Trek. The crew was then able to cross the galaxy during the TV commercials.

Dear Merlin,

I recently read a quote by Oscar Wilde that referred to "seven heavens." What are the seven heavens?

CURTIS LAZARZ

ST. CLOUD, FLORIDA

The seven heavens are a sequence of ultimate spiritual bliss based on verses from the Koran, the Muslim religious text. Allah is said to have created seven heavenly concentric spheres around Earth, with the seventh heaven being the farthest and largest sphere, which contains all the stars as well as Allah's dwelling place.

In common parlance, of course, to be in seventh heaven refers to being in a very pleasing state of mind.

Dear Merlin,

What is antimatter? Does it exist naturally or is it manmade?
SHORTI THOMPSON
BEDFORD, VIRGINIA

*A*ntimatter is *real*. That favorite fuel of science-fiction spacecraft is matter that is composed of anti-particles. In fact, every "normal" particle has a corresponding anti-particle that is identical in mass but has the opposite value for various physical properties, including charge. For example, the electron's anti-particle is the positively charged anti-electron— commonly called the positron.

When a particle meets its anti-particle, the encounter is swift and bright. They annihilate each other by turning 100 percent of their mass into energy. Indeed, part of the Sun's total energy budget is derived from electron-positron annihilation immediately after positrons are formed as a byproduct of thermonuclear fusion in the Sun's core.

Anti-particles are routinely created in particle accelerators. But in the universe, freely roaming anti-particles are rare, and fully bound atoms of antimatter have never been found. If you worry that one day you may meet aliens who are composed of antimatter, you can test your hypothesis quite easily. Just toss them some matter—anything from your pocket will do. If they explode instantly, your antimatter suspicions are correct.

Dear Merlin,

Television broadcasts, from what I have heard, are not confined to just our Earth but continue on into the far reaches of space. If we were to receive television images from some distant planet, could we deduce how its inhabitants compare to us in relative size as well as draw some conclusion about the nature of their planet or civilization?

Conversely, if our broadcasts are being received by another planet, I wonder if they are studying us or trying to devise methods to filter out The Simpsons *and* Beavis and Butt-Head.

MIKE HATCHIMONJI

LA PALMA, CALIFORNIA

*I*t is not clear whether the existence of television actually correlates with the existence of civilization.

But to the extent that it does, there remains the risk that a single TV show or sitcom episode will form the basis for the analysis of an entire culture—a scary thought indeed. Anthropologists are often faced with this sort of dilemma. Yet armed with enough examples, cosmic anthropologists (cosmopologists?) would clearly learn a substantial amount about the mental state of couch potatoes.

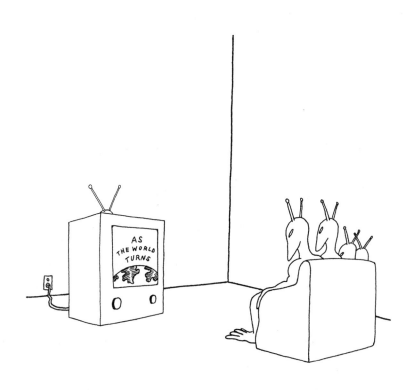

Dear Merlin,
    What is the probability of an individual coming into contact with the same molecule of air more than once in a lifetime?
    ROBERT L. BRINKLEY
    DAYTON, OHIO

If you never leave Dayton, Ohio, your chances are quite good.

Otherwise, if we assume that Earth's atmosphere stays well mixed and if we assume you live to be ninety-eight years old, the chances are about one in 6,000,000,000,000,000 (6 quadrillion) that you will breathe the same outdoor air molecule twice in your lifetime.

Incidentally, this is the same probability that you will breathe an air molecule that was also breathed by the oil tycoon John David Rockefeller, or anybody else who has lived for ninety-eight years.

Dear Merlin,
   What is the unified field theory?
   SHORTI THOMPSON
   BEDFORD, VIRGINIA

*O*nce upon a time there were three known forces in the universe: electricity, magnetism, and gravitation. In the late nineteenth century, primarily through the work of James Clerk Maxwell and Heinrich Rudolf Hertz, magnetism and electricity were found to be different aspects of the same phenomenon. These two forces were unified to become the electromagnetic force.

In the first quarter of the twentieth century, the atom was probed to unprecedented detail. Two additional forces were revealed, which came to be known as the strong and weak nuclear forces.

In the spirit of nineteenth-century physics, twentieth-century physics has devoted much attention to combining the four remaining forces into one common description, on the (possibly misguided) assumption that only one basic force is required to understand all of nature. In 1978, the Nobel prize in physics was awarded to Abdus Salam, Steven Weinberg, and Sheldon Glashow for unifying the weak nuclear force with the electromagnetic force to make the electroweak force.

And then there were three again. A unified field theory would combine the electroweak force with the strong nuclear force into an as yet untitled entity. This remains to be accomplished.

And then there would be two. To combine gravity with the other forces requires what is now called a grand unification theory. This also remains to be accomplished.

Dear Merlin,

I enjoy your comments, especially the sarcasm. (My favorite author is Mark Twain.) Being a doddering old fool, I also take great delight in needling humans, mainly politicians. But I met my match when I took on you—an alien.

Actually, since I've lived my three score and ten, I think I might like to take a trip to a black hole just to see what they are made of. To pay my way, I'll take along a rocketload of preachers and cosmologists. Perhaps you would like to go along as my navigator?

ROBERT H. DUNGAN

ARBOLES, COLORADO

*F*alling into a black hole requires very little navigational talent—you won't need Merlin's help. Before you go, however, please leave behind at least one preacher and one cosmologist. After your body is ripped apart by the black hole's tidal forces, your eulogy may require the services of both.

Dear Merlin,

I have read that at least some black holes are drawing in material from nearby stars. Will they ever get filled up?

GEORGE S. ROBINSON

DESTREHAN, LOUISIANA

Black holes grow in direct proportion to the mass consumed. For example, if a black hole were to double its mass by dining on a nearby red giant, then its radius would double in response. In other words, a black hole can never get full. It just keeps growing and growing and growing and . . .

Unlike most people, however, a black hole never fills up from having eaten too much. And since black hole food always makes a one-way trip, there is no risk of black hole throw-up.

Dear Merlin,

Did you make a common human error in your answer concerning the growth of black holes? If the mass of the black hole doubled, I suggest that the radius would only increase by the cube root of two. This is true for real objects, at least.

ROBERT H. DUNGAN

ARBOLES, COLORADO

By convention, the radius of a black hole is the radius of its event horizon, which is that high-gravity region of space within which light cannot escape. The material object that became a black hole actually continued to collapse down to a point after its surface passed through its own event horizon.

Applying laws of energy and gravity and some simple algebra, you can derive the precise relationship between the mass of a black hole and its radius. If you must know: $R = (2G/c^2)\, M$, where R is the radius of the event horizon, G is Newton's gravitational constant, c is the speed of light, and M is the mass of the black hole. Notice from the equation that the mass and radius vary in lockstep: if M doubles or triples, then so must R.

P.S. If you ever were to fall into a black hole, you would agree that it was a real object.

Dear Merlin,

Could there be another side or another dimension to a black hole? If so, when stars get captured in it, would there be so much pressure in between the two sides that the star would implode and make the black hole's gravity even stronger? Could this be the reason why X-rays are detected from black holes rather than from gas swirling around them?

KARL ENGELMANN

MISHAWAKA, INDIANA

*N*ot likely. Current evidence strongly supports the swirling-gas, accretion-disk model for X-rays emitted around neutron stars and black holes.

It appears that you like speculative theories. In the early 1970s it was demonstrated that mathematically there exists a negative black hole—one that regurgitates matter rather than eats it. The name "white hole" seemed appropriate. To carry these mathematical hijinks further, every black hole was connected to a white hole through a tunnel-like worm hole. Remarkably, none of this is in conflict with Einstein's general theory of relativity. Unfortunately, no object ever observed is best explained by a white-hole model.

Dear Merlin,

Are black holes gathering matter in preparation for another big bang in another time and dimension?

HAROLD W. CURTIS

BEVERLY HILLS, FLORIDA

What black holes do on the other side of their event horizon is nobody else's business.

The contents of a black hole are outside of our interacting universe and are therefore unknowable. It makes fertile ground, however, for science-fiction stories. Why don't you write one?

Dear Merlin,

What is the possibility of other big bangs in the universe at enormous distances beyond the most distant known quasar?

FRANK DESIMONE

STATEN ISLAND, NEW YORK

Current evidence suggests that the universe was born in a single big bang.

If you long for other big bangs, you will be happy to know that we cannot rule out the existence of other universes outside our own. Perhaps our universe is one of many contained within a mega-universe where mini-universes are born all the time. And perhaps each universe has its own array of physical laws that differ from ours.

Unfortunately, at least for now, such exotic ideas come closer to being science fiction than science fact.

Dear Merlin,

    What chance is there that intelligent life exists on other planets?
ZANE BELL
WILLIAMS, OREGON

The most successful attempt to estimate this likelihood was made by the American astronomer Frank Drake of the University of California at Santa Cruz. He developed what has come to be known as the Drake equation, which is a sequence of fractions that when multiplied together estimate the number of intelligent civilizations in the universe.

The equation includes, among other terms, the fraction of stars in the universe that have planets, the fraction of planets that are habitable for life as we know it, the fraction of habitable planets that actually do have life, the fraction of planets with life that have intelligent life, and the fraction of planets with life that is intelligent enough (and has the technology) to communicate with us.

The universe has over a sextillion (10^{21}) stars. While there is large uncertainty in the values chosen for each fraction, most estimates lead to the inescapable conclusion that earthlings are not alone in the universe. And if you want proof, just visit Omniscia, Merlin's home planet in the Draziw star system of the Andromeda galaxy.

XII

THE WORLD
of MERLIN

Dear Merlin,

I read in The Once and Future King *that you are younging instead of aging like the rest of us. Is this true? Shall I save my baby rattles for you?*

RON KENNEDY (AGE 85)
SCOTTSDALE, ARIZONA

You are probably not the only Ron on Earth, so do not be disappointed that there are at least two Merlins in the universe.

Please keep your baby rattle, or give it to somebody else, because this Merlin lives forward in time, has a very good knowledge of the past (especially the historical scientific endeavors of the human species), and does not predict the future.

Dear Merlin,

    Since your home galaxy, Andromeda, is over 2 million light-years away and nothing travels faster than the speed of light, why did you leave home to become a visiting scholar on Earth long before modern man appeared on the scene?

    JASON KAMINSKY

    ST. LOUIS, MISSOURI

    *M*erlin was born on Omniscia 4.6 billion years ago, which coincides with Earth when it was forming in the Milky Way galaxy. It was obvious billions of years ago that conditions on Earth were ripe for the emergence of intelligent creatures—whales, humans, etc.

Dear Merlin,

I read with interest your prescient decision to make the trip to planet Earth to become a visiting scholar. Was there really no better or closer place than Earth?

BILL PARKINS

WOODLAND HILLS, CALIFORNIA

$\mathcal{M}$erlin has many omniscient cousins who are visiting scholars on other developing worlds. Merlin finds humans, however, to be quite interesting. They have a rich history of scientific thought and they ask intriguing questions. Merlin is pleased to assist in the noble quest of achieving scientific literacy among all humans.

Dear Merlin,

How often do you visit Omniscia? Have Omniscians learned anything new recently?

LT. COL. ROBERT A. FOWLER
SAN ANTONIO, TEXAS

$\mathcal{M}$erlin visits home about every thousand Earth years.

Many scholars of Omniscia, Merlin included, seek to study other civilizations. Merlin's mission on Earth is to observe the discovery of scientific knowledge and to help disseminate this knowledge across the land. Merlin is, however, a passive observer and cannot risk altering the natural course of Earth history by sharing with Earth residents the scientific discoveries of other civilizations, including those of Omniscia.

Dear Merlin,

Do you have a family on Omniscia?

Lauren and Richard Vosburgh

New Milford, Connecticut

Yes, Merlin has a large family and many relatives. On Omniscia, one chooses a mate for life, and the older of the two is the one who is genetically encoded to bear offspring.

On Omniscia, however, there is no gender. And everyone's skin reflects photons with a wavelength of about 4,800 angstroms (Omniscians appear blue). A tranquil consequence is that the color- and gender-based disagreements that are common among Earth humans do not exist on Omniscia.

Dear Merlin,
  Enclosed is a tiny calendar as a gift from us. Printed on the upper half is

From the Demings
ALICE & BOB
Y'all come to see us
Fredericksburg, Texas.

Just peel off the back and you can stick it on the dashboard of your car.
ALICE AND BOB DEMING
FREDERICKSBURG, TEXAS

*T*hank you, Alice and Bob. Your hospitality is befitting of your residence in the state of Texas. Unlike most Texans, however, Merlin does not drive a car. Merlin has other ways of getting around the galaxy, none of which don a dashboard.

Dear Merlin,
   What is your mission on Earth?
   ARNOLD LIVINGSTON
   LOS ANGELES, CALIFORNIA

$\mathcal{M}$erlin is just visiting this planet. While here, Merlin intends

1. To be friendly and conversational.
2. To attract cosmic queries from all interested readers.
3. To help enlighten Earth residents about their rich scientific past and present.
4. To introduce readers, on special occasions, to some of Merlin's friends.
5. And to promote truth, justice, and the cosmic way.

SOME OF MERLIN'S FRIENDS AND ACQUAINTANCES

Ångström, Anders Jonas, 1814–1874, Swedish physicist.

Considered one of the founders of spectroscopy, the study of light after it has been broken up into its component colors by a prism or other dispersing device. Ångström made seminal contributions to the study of atomic spectra and the study of the Sun's spectra, where he inferred the presence of hydrogen. A unit of wavelength known as the angstrom unit (1 Å = 10^{-10} meters) is named in his honor. With this measure, the visible light spectrum spans from blue at about 4,000Å to red at about 7,000Å.

Archimedes, c. 287–212 B.C., Greek mathematician.

One of the greatest mathematicians of ancient times. He developed rigorous techniques to measure curved lines, area, and solids. Also well known for his work with screws, levers, and the buoyancy of solid objects.

Barnard, Edward Emerson, 1857–1923, American astronomer.

An ace observer. In 1916, Barnard discovered a star, now named in his honor, that is the second closest star system to the

Sun. Its motion through space across our field of view is so fast compared with other stars that from one year to the next its position in the sky changes by 10 arc seconds, one of the largest on record. Over 180 years, Barnard's star moves the diameter of the full moon. Ten thousand years from now it will have become the Sun's nearest neighbor and will be moving across the sky at about 25 arc seconds per year.

Bradley, James, 1693–1762, English astronomer.

Another ace observer. Bradley made many attempts to measure the shift of a star's position against the background stars as Earth moves from one side of the Sun to the other. A successful measurement would enable a reliable distance to be determined. Although failing to make such a detection, he noted an offset in all stellar positions that depended on whether Earth, in its orbit, was moving toward, away from, or sideways to the star. He correctly attributed this effect to the finite speed of light.

Browning, Robert, 1812–1889, English poet.

Merlin has never met the chap, but overheard him murmuring beautiful phrases about life and death in an English pub.

Cavendish, Henry, 1731–1810, English physicist.

As a scientist, Cavendish is known for measuring the density of Earth and for demonstrating the existence of hydrogen ("inflammable air") as a distinct element. Cavendish's name, however, is far more famous from its association with the Cavendish Physics Laboratories in Cambridge, England, where much of twentieth-century atomic physics was born.

Celsius, Anders, 1701–1744, Swedish chemist.

Though an astronomer by training, Celsius is best known for the centigrade temperature scale that he heavily promoted and that bears his name.

Cherenkov, Pavel Alekseevich, 1904–1990, Russian physicist.

Pioneered the study of light moving through transparent things like glass and water. In these and other mediums, the speed of light is reduced from its speed in a vacuum. He noted that when a particle in this medium is accelerated to speeds faster than the speed of light in the medium, the particle emits the light-wave equivalent of a sonic boom. This emission was named Cherenkov radiation in his honor.

Claus, Santa, A.D. 4th Century, formerly Saint Nicholas.

Patron saint of children, pawnbrokers, unmarried girls, and sailors. The last time Merlin checked, Mr. and Mrs. Claus still resided at the chilly North Pole of Earth's rotation axis.

Copernicus, Nicolaus, 1473–1543, Polish astronomer.

Good friend of Merlin's. Nick, the greatest astronomer since Hipparchus, is generally credited with restoring the idea of a Sun-centered (heliocentric) universe. For nearly 1,500 years, the Earth-centered (geocentric) universe had been predominant.

Doppler, Christian, 1803–1853, Austrian physicist.

Merlin never met Herr Doppler. A distinguished physicist in his day, Doppler is best known for the spectral effect that bears his name. When an object in motion emits a wave of anything (e.g., sound or light), the natural frequency of the wave is increased if the object is approaching you and decreased if the object is receding. This general principle, the Doppler shift, reveals itself in countless ways, including train whistles, race cars, and the expanding universe.

Drake, Frank Donald b. 1930, American astronomer.
Best known for his work on the search for extraterrestrial intelligence, and in particular, for an equation that bears his name. The Drake equation is a means to calculate the probability of there being other intelligent and technologically able life forms in the galaxy.

Eddington, Sir Arthur, 1882–1944, English astrophysicist.
Eddington took his tandem knowledge of physics and astronomy and married them in his research to become the first astrophysicist. A brilliant scientist with a tireless interest in the latest and most important problems of the day, Sir Arthur is well known for conducting the first confirming measurements of the curvature of space-time as predicted by Albert Einstein in his modern theory of gravity, which is better known as the general theory of relativity. Sir Arthur also attempted to deduce the nature of stars and other cosmic phenomena from physical principles. While not always correct, his efforts reliably stimulated further research by others.

Einstein, Albert, 1879–1955, German-American physicist.
A good friend of Merlin's, Einstein made contributions to our physical understanding of the universe that are rivaled only by those of Isaac Newton. In 1905, Al proposed his revolutionary special theory of relativity, where space and time are conjoined. This conceptual framework allowed him to make counterintuitive

predictions about the mass, the flow of time, and the physical dimensions observed of an object as its speed approaches the speed of light. In his 1916 theory of gravity—the general theory of relativity—Albert interpreted gravity as the curvature of space-time through which matter falls, rather than as a conventional force that acts at a distance. To date, all reliable experiments that have ever been conducted have confirmed the predictions of relativity.

Eliot, Thomas Stearns, 1888–1965, American-born English poet.

In one of his poems, he seemed to know how the universe would end long before astrophysicists figured it out.

Euler, Leonhard, 1707–1783, Swiss mathematician.

Unquestionably one of the most brilliant and prolific mathematicians of all time. Merlin once overheard the French physicist-astronomer François Arago say of Euler, "He calculated just as men breathe, as eagles sustain themselves in the air." Merlin agrees.

Fahrenheit, Gabriel Daniel, 1686–1736, Polish-born German physicist.

Not unlike Thomas Edison in his approach to experimenting, Fahrenheit, without being driven by a particular scientific principle or a motivating theory, spent much of his scientific career tinkering with various liquids and glass materials in an attempt to design a reliable thermometer. What emerged from his tireless efforts was the mercury thermometer and a temperature scale named in his honor.

Foucault, Jean Bernard Léon, 1819–1868, French physicist.

An all-around smart fellow, Foucault derived the first very reliable value for the speed of light (over 99 percent accurate), demonstrated that light travels more slowly in water than in air, and invented Foucault's pendulum, which demonstrates the rotation of Earth in space—among his many achievements.

Galilei, Galileo, 1564–1642, Italian physicist.
   Merlin and Galileo go way back. While not the inventor of the telescope, Galileo may have been the first to look up with it. What lay before him was a garden of cosmic knowledge that permanently altered the landscape of scientific thought. His discoveries ranged from simple observations that the Moon's surface is not smooth (as presupposed) to the fact that Earth cannot be the center of all motion, since Jupiter has a set of moons all to itself. His findings and his relentless ego got him in trouble with the Catholic Church. He was found guilty of heresy and, to avoid torture, was forced to sign a confession that renounced his data. Galileo (rather, his corpse) was found innocent of all charges somewhat later (350 years) by Pope John Paul II.

Gamow, George, 1904–1968, Russian-American physicist.
   A distinguished physicist at a time when human understanding of the atom and the universe took fundamental leaps. In 1946, Gamow proposed what came to be known as the big-bang model of the universe, which came with testable predictions of the abundance of heavy elements and of a background remnant of microwaves from the original explosion.

Hertz, Heinrich Rudolf, 1857–1894, German physicist.
   Demonstrated that radio waves are just another form of electromagnetic waves, akin to visible light, thus enabling the intellectual unification of previously disjoint forms of energy. The familiar unit of electromagnetic frequency is named in his honor.

Hipparchus, c. 146 B.C., Greek astronomer, mathematician, geographer.

Invented the longitude and latitude system of spherical coordinates, which remains in use today. Hipparchus made one of the earliest reliable catalogues of stellar positions in the sky. In so doing, he noticed the wobble of Earth's axis, otherwise known as the precession of the equinoxes. So thorough and accurate was his catalogue that Edmund Halley (of Halley's comet fame) referenced it 1,800 years later.

Homer, 9th century B.C., Ionian poet.

The *Iliad* and the *Odyssey* are two epic works attributed to him. In the *Iliad*, Homer claims some familiarity with Orion, the hunter.

Hubble, Edwin Powell, 1889–1953, American astronomer.

Among many seminal contributions to observational astronomy, Hubble discovered the expanding universe in 1929. Law and boxing were two early interests before he turned to the heavens. Merlin and Hubble were chums.

Humanson, Milton Lasell, 1891–1972, American astronomer.

Best known for his work on the hundred-inch telescope at the Mount Wilson Observatory, and later on the two hundred-inch telescope at Mount Palomar, Humanson obtained spectra of galaxies in the late 1920s and for decades to follow. His data enabled Edwin Hubble to extend the discovery that distant galaxies recede faster than nearby ones.

Kant, Immanuel, 1724–1804, German philosopher.

Among astrophysicists, Kant is best remembered for having proposed in a 1755 essay the "nebular hypothesis" to explain the origin of the solar system. Kant suggested that a large spinning gas cloud had flattened as it collapsed under its own gravity. A large central nucleation had formed the Sun while smaller nucleations had formed the planets. While various modifications to this suggestion have been required over the years, the basic idea and scenario are correct. Extending this idea to the entire galaxy, Kant also supposed that the fuzzy "stars" in the sky were other galaxies—island universes distinct from our own—an idea later confirmed by Edwin Hubble in 1929.

Kepler, Johannes, 1571–1630, German mathematician and astronomer.

Kepler proposed the first truly predictive mathematical theory of the universe through his laws of planetary motion. Isaac Newton later showed that Kepler's laws are derived easily from more basic theories of gravity.

La Bruyère, Jean de, 1645–1696, French writer.

Merlin never met Mr. La Bruyère but one day overheard in a Paris *jardin* his ruminations on the meaning of life and death.

Laplace, Pierre Simon, 1749–1827, French mathematician and astronomer.

Duly famous in the annals of astronomy for many reasons. Most notably, Laplace updated Isaac Newton's laws of gravity to allow for the hard-to-predict multiple effects of many sources of gravity acting simultaneously. In what is today called perturbation theory, Laplace's technique allowed one to calculate planetary orbits with unprecedented precision. In the face of this enlightened understanding of celestial motions, Merlin overheard Napoleon Bonaparte comment to Laplace that there was no mention of God in his book, whereupon Laplace replied, "Sir, I have no need of that hypothesis." Laplace also postulated the existence of an object with such high gravity that light might not escape, and he proposed independently (and somewhat later than Immanuel Kant) that the system of planets owes its origin to a collapsing, flattening, rotating gas cloud.

Leibniz, Gottfried Wilhelm, 1646–1716, German mathematician.

A contemporary of Isaac Newton's, Leibniz independently invented the powerful mathematical technique known as calculus. Calculus to Leibniz was a mathematical curiosity, whereas calculus to Newton was an indispensable tool with which one could decode the motions of celestial objects.

Lippershey, Hans, c. 1570–1619, Dutch optician.

Credited with being the first person to assemble two lenses in such a way that objects appear closer to the person who looks through them. This invention is better known as the telescope.

Lord Kelvin. See Sir William Thomson

Mach, Ernst, 1838–1916, Austrian physicist and philosopher.

Mach's writings on the fact that space and time have meaning only when they refer to observable relations between objects led Albert Einstein to propose his special theory of relativity in 1905.

Mann, Horace, 1796–1859, American educator.

Never met him, but Merlin once overheard him reflect upon one's responsibility to others in life.

Maxwell, James Clerk, 1831–1879, Scottish physicist.

Maxwell made fundamental contributions to human understanding of light. In the equations that bear his name, all of the classical (prerelativity and pre–quantum mechanics) description of electromagnetism is elegantly captured. Maxwell is also celebrated for his contributions to the behavior of molecules in a gas.

Messier, Charles, 1730–1817, French astronomer.
Messier is famous for his list of more than one hundred fuzzy objects in the nighttime sky, conceived as a list of "not comets." Messier's incentive was to prevent comet hunters from being confused during their search. Of course, bad optics can make everything look fuzzy. Indeed, Messier's telescope may have been of dubious quality, because many objects on his list are loose assemblies of stars that could not possibly have been confused with a comet by a comet hunter who was armed with good optics.

Michelson, Albert, 1852–1931, American physicist.
Best known for his development of the interferometer, which is an extremely sensitive optical device that can be used to measure, among other things, the speed of light to unprecedented precision. Teamed with Edward Morley in 1887, he demonstrated that the speed of light was independent of the direction that Earth moved through the ether, thus casting serious doubt on ether's existence as a medium through which light must travel. In 1907, Michelson was the first American to receive the Nobel prize in physics.

Minkowski, Hermann, 1864–1909, Lithuanian mathematician.
A pure mathematician who is best remembered for laying some of the mathematical foundations of space-time for Albert Einstein's general theory of relativity.

Morley, Edward Williams, 1838–1923, American chemist.
See ALBERT MICHELSON

Newton, Isaac, 1642–1727, English physicist.
An often-quoted comment from Isaac Newton proclaims that if he can see farther than other men, it is because he stands upon the shoulders of giants who came before him. This may indeed have been true (especially if the giants are Copernicus, Kepler, and Galileo), but the real secret to his distant vision might simply have been that he was surrounded by intellectual midgets. A candidate for one of the greatest intellects ever to walk the Earth, Isaac is Merlin's best friend in all of time. In spite of this, he spent his most scientifically productive years alone. Over that time he discovered the laws of gravity and many laws of optics. He also invented calculus. A statue of him at Trinity College in Cambridge, England, proclaims: "Of the human species, the brilliance of Isaac Newton reigns supreme." Merlin agrees.

Olbers, Heinrich Wilhelm Matthäus, 1758–1840, German astronomer.
Popularized the idea that if the universe were infinite, the collective energy from the stars that occupy the ever-increasing volumes of space would render the night sky ablaze with light. That we live in an expanding universe, which had a beginning, handily resolves this paradox, because starlight loses energy faster than it sums together with the light of other stars.

Pogson, Norman R., 1829–1891, English astronomer.
Pogson noted that the brightest and dimmest stars in the nighttime sky had about a factor of one hundred difference in brightness. Knowing that the ancients had assigned a brightness magnitude scale of one to six for all stars, Pogson recommended

that the brightness of all stars be codified into a formal mathematical relation where a brightness factor of the fifth root of one hundred separates consecutive magnitudes. Sometimes in the world of measurement, familiarity is more important than ease of use. Pogson's arcane system remains in use today among most observational astronomers.

Rex, Thelma, 65 million B.C., Prehistoric tyrannosaur.

Eyewitness to the asteroid collision with Earth that led to the extinction of many of the world's species, including all dinosaurs. Thelma and Merlin were buddies.

Rømer, Ole Christensen, 1644–1710, Danish astronomer and mathematician.

Rømer noticed that eclipses of Jupiter's moon occurred about one thousand seconds sooner than predicted whenever Earth, in its orbit, was closest to Jupiter, compared with when Earth is at its farthest. He correctly attributed this delay to a limit on the speed of light.

Slipher, Vesto Melvin, 1875–1969, American astronomer.

An accomplished astronomer who is best known for having obtained spectra of spiral nebulas that enabled Edwin Hubble to conclude that the spiral nebulas were indeed entire galaxies external to our own Milky Way, and that nearly all were moving away from us.

Thomson, William (Baron Kelvin of Largs), 1824–1907, British physicist.

A precocious lad, William Thomson graduated from the University of Glasgow at age ten. He became a major contributor to human understanding of the electromagnetic force and the study of thermal energy better known as thermodynamics. The Kelvin absolute-temperature scale, where zero degrees is the coldest possible temperature, is named in his honor.

Trog & Lodyte, 10,000 B.C., cave dwellers (married).
Good friends of Merlin's. They discovered the wheel one day when they were bringing home the groceries. The task was made much easier by placing the groceries on a plank and rolling the plank along logs. No doubt this was the world's first shopping cart.

Voznesensky, Andrei, b. 1933, Russian poet.
One of few poets to tap scientific terms to convey poetic imagery.

Wallis, John, 1616–1703, English mathematician.
One of the earliest people to tackle the notion of infinity. He introduced the infinity symbol, ∞, which remains in use today. Out of this effort emerged a facility with infinite series of numbers, which when added, summed, divided, or multiplied yield unique numerical quantities. One such series was a formula for pi.

Weinberg, Steven, b. 1933, American physicist.
A theoretical particle physicist, he shared the 1979 Nobel prize in physics with Sheldon Glashow and Abdus Salam for theoretically combining the weak nuclear force and the electromagnetic force into a single description and thus "unifying" these two forces of nature, fueling the quest for a unified theory of all known forces.

Wright, Wilbur, 1867–1912; Orville Wright, 1871–1948, American inventors.

Using a wind tunnel as the testing ground for over two hundred airplane designs, the Wright brothers finally put to test their *Wright Flyer* at Kitty Hawk, North Carolina, in 1903. While not on the historic first flight, Merlin was in attendance and went joyriding with each of them on subsequent trips.

SUGGESTED READING
Magazines

Star Date
2609 University, Room 3.118
University of Texas at Austin
Austin, TX 78712

Feature articles on the universe, monthly star charts, and of course the original Merlin column make this magazine ideal for the weekend astronomer and anyone who wants to stay informed.

Science News
Science Service
1719 N Street NW
Washington, DC 20036

All-purpose science reporting by intelligent writers who care that they get their facts straight. Appearing weekly, *Science News* is the easiest way to stay current and conversational in all the major sciences.

Scientific American
415 Madison Avenue
New York, NY 10017
The latest science written by the experts in the field. Not always as readable as it claims, or as it wishes it were, *Scientific American* remains the most authoritative science magazine intended for the lay reader.

Discover
114 Fifth Avenue
New York, NY 10011
Heavily illustrated articles in flashy areas of science make *Discover* one of the most popular science magazines, even if it occasionally blurs the distinction between fringe and mainstream science.

Sky and Telescope
Sky Publishing Corporation
49 Bay State Road
Cambridge, MA 02138
Unimpeachable as the monthly bible for the serious amateur astronomer. If you don't know your magnitude scale or your Mizar from your Alcor, then you risk being intimidated.

Astronomy
Kalmbach Publishing Co.
21027 Crossroads Circle
P.O. Box 1612
Waukesha, WI 53187
Astronomy not only holds its own against *Sky and Telescope*, it has historically enjoyed a larger circulation. Credit the emphasis on

reader-friendly content and seemingly endless photo spreads of cosmic splendor.

Books

Merlin's Tour of the Universe
Neil de Grasse Tyson
Mainstreet Press/Doubleday, New York, 1997
(originally published by Columbia University Press, 1989)
Merlin's first Q&A journey through cosmic knowledge.

The New York Times Book of Science Literacy:
What Everyone Needs to Know, From Newton to the Knuckleball
Richard Flaste, ed.
HarperCollins, New York, 1992
Straight questions and answers on the fundamental science topics of modern times. An important first step toward science literacy.

As playful as a question-and-answer format can be, genuine encyclopedic research has no substitute. If you seek further cosmic enlightenment and you buy only one book, try

Cambridge Encyclopedia of Astronomy
Jean Adouze, ed.
Cambridge University Press, New York, 1992
It contains everything you ever (and never) wanted to know about the universe in an attractive coffee-table-book size. But unlike other coffee-table books, this one does not use pretty pictures as a substitute for content.

Some Historical References

Biographical Dictionary of Scientists
Trevor Williams, ed.
HarperCollins, New York, 1994

Asimov's Chronology of Science and Discovery
Isaac Asimov
Harper & Row, New York, 1989

The Timetables of History
(based on Werner Stein's *Kulturfahrplan*, 1946)
Bernard Grun
Touchstone/Simon & Schuster, New York, 1991

A Source Book in Astronomy and Astrophysics, 1900–1975
Kenneth R. Lang and Owen Gingerich, eds.
Harvard University Press, Cambridge, Mass., 1979

The Facts on File Encyclopedia of the 20th Century
John Drexel, ed.
Facts on File, New York, 1991

INDEX

NEIL DE GRASSE TYSON was born and raised in New York City where he was educated in the public schools through his graduation from the Bronx High School of Science. He went on to earn his B.A. in physics from Harvard and his Ph.D. in astrophysics from Columbia. His popular monthly column "Universe" appears in *Natural History* magazine, and he holds a joint position as visiting research scientist at Princeton's Department of Astrophysics and as the Frederick P. Rose Director of the Hayden Planetarium, where he is project scientist for the reconstruction of one of the nation's greatest astronomical attractions.